Competitive Edge

Contributors

M. Therese Flaherty
Hiroyuki Itami
John G. Linvill
Daniel I. Okimoto
Takuo Sugano
Michiyuki Uenohara
Franklin B. Weinstein

Competitive Edge

THE SEMICONDUCTOR INDUSTRY
IN THE U.S. AND JAPAN

Edited by Daniel I. Okimoto, Takuo Sugano,
and Franklin B. Weinstein

STANFORD UNIVERSITY PRESS   1984
Stanford, California

ISIS Studies in International Policy

Sponsored by the Northeast Asia–United States Forum on International Policy
of the International Strategic Institute at Stanford

Stanford University Press
Stanford, California
© 1984 by the Board of Trustees of the
Leland Stanford Junior University
Printed in the United States of America
ISBN 0-8047-1225-5
LC 83-40107

TO JOHN W. LEWIS

in grateful appreciation of the vision and leadership
that sustain the Northeast Asia–U.S. Forum

Preface

This book is the product of a collaborative effort to assess the dynamics of the intensifying competition between the U.S. and Japan in the semiconductor industry. The semiconductor competition opens a new era of competition in a broad area of traditional American leadership—high technology. If this competition degenerates into confrontation, the results could be even more troublesome than previous clashes over consumer electronics, automobiles, and steel. The intrinsic importance of this sector, and the complexity of the issues defining the competitive relationship between the U.S. and Japanese semiconductor industries, suggest the need for an objective study that incorporates both U.S. and Japanese perspectives.

The potential for misunderstanding and conflict is great. In the late 1970's, U.S. semiconductor manufacturers began to view with alarm the rapid growth of Japan's market share, particularly with respect to the then-dominant chip, the 16K RAM (a random access memory containing approximately 16,000 cells). Many felt that the Japanese were competing unfairly. The Japanese were accused of sending "spies" to gain access to the secrets of semiconductor companies in California's "Silicon Valley." There were allegations that the Japanese were dumping 16K RAMs in an effort to gain market share. The Japanese were said to enjoy government subsidies and assured access to low-cost capital that their U.S. competitors could not match. Furthermore, they had unimpeded access to U.S. markets, whereas American companies had to overcome a variety of barriers if they wished to sell or manufacture in Japan.

Underlying this perception of the Japanese as unfair competitors was the belief that the Japanese semiconductor makers operated as part of a Japanese monolith, with companies, banks, and government agencies working together in ways that would have been considered unethical and, in some cases, illegal, if pursued in the United States.

For their part, Japanese semiconductor makers expressed deep concern at what they saw as unfounded allegations by U.S. companies frustrated by their inability to compete. They felt that American companies were in large measure responsible for whatever problems they were experiencing in both Japan and the United States. The Japanese learned the English language, studied American techniques, analyzed the U.S. market, adapted, and worked hard to sell. The Americans, it was said, had not expended comparable efforts to learn how to sell in Japan. Furthermore, the Japanese claimed that their chips had attained superior levels of quality, partly because the Japanese buyer demanded it and partly because of a Japanese decision to rely more heavily on automation; the Americans had chosen to make greater use of cheap labor for assembly in the Third World. Japanese penetration of the U.S. market for 16K RAMs was facilitated, according to the Japanese, not by dumping on their part but by the inability of U.S. companies, which had slowed investment during the recession of the mid-1970's, to meet the exploding demand for chips in the latter years of the decade.

The Japanese acknowledged government support for research and development, but they claimed that this paled in comparison with the backing that U.S. defense and space programs had provided at critical points in the U.S. industry's development. Moreover, if restrictions imposed by the American system made it difficult for U.S. companies to compete, let the Americans adapt. Why, the Japanese asked, should they be penalized for success?

Finally, the Japanese noted that the U.S. companies still controlled about two-thirds of the world market for semiconductor products, and that U.S. semiconductor exports to Japan, overall, exceeded the flow in the other direction. The Americans, accustomed to unchallenged domination of the market, were said to be overreacting to the emergence of their first serious competitive challenge.

A number of studies touching on U.S.-Japanese competition in the semiconductor industry have been completed in the last few years. These include useful works by the U.S. Department of Commerce, U.S. International Trade Commission, U.S. Department of State, Office of Technology Assessment, Charles River Associates, Chase Manhattan Bank, and a team of scholars at the University of California, Berkeley. None of these studies involved Japanese participation, however, nor does any offer a systematic analysis of Japan and its semiconductor industry. Most of them treat only the U.S. industry in depth. Above all, none of those studies provides a comparative analysis of the systemic factors—political, economic, and technological—that influence the ability of each country's semiconductor industry to compete successfully. That is the task we have set for ourselves in this book.

This study was undertaken by a binational working group that included economists, electrical engineers, political scientists, and sociologists. In April 1980, the Japanese members of the group visited Stanford to meet with their U.S. counterparts and with an invited group of U.S. semiconductor industry executives, Japanese executives resident in the United States, government officials from both sides, and others with a special interest in the topic. The purpose of that meeting was to review existing work on the subject and to define the problems on which our own study should focus. We decided at that time to concentrate on the basic technological, financial, and political factors affecting the competitive relationship between the two industries, eschewing a detailed consideration of some of the more visible, but probably ephemeral, points of controversy. During the remainder of 1980, the Japanese researchers visited the United States, conducting interviews with U.S. semiconductor industry executives in Dallas, Phoenix, and California's Silicon Valley. American members of the working group went to Japan to interview Japanese executives and government officials.

At meetings held in Tokyo in December 1980–January 1981, outlines of the chapters were discussed. Further field research was undertaken, and a preliminary draft of the report was presented for discussion at a meeting held at Stanford in May 1981. At that meeting, a great deal of valuable commentary was provided by industry representatives invited to critique the draft chapters. The tentative findings of the group were also discussed at public meetings held in Palo Alto, New York, and Washington. Those meetings, organized by the Japan Society with financial support from the Japan–United States Friendship Commission, yielded valuable comments. Additional research was done, successive revisions were undertaken in response to further comments, and, in the spring of 1983, the final version was completed.

In addition to the coauthors of this book, members of the working group included Everett M. Rogers, David Gibson, and Moriaki Tsuchiya, who collaborated on a paper dealing with management practices in U.S. and Japanese semiconductor companies; Marcy Avrin and Michael Boskin, who compared national tax systems, particularly in terms of their effects on capital investments; and W. Edward Steinmueller and Yuji Masuda, who examined the two industrial structures, with respect especially to vertical integration. All members made valuable contributions to our collective understanding. Many of their ideas have been incorporated into the chapters of this book. We feel extremely fortunate to have had the stimulus of their participation in our project. A special word of thanks is due to W. Edward Steinmueller, whose vast knowledge of the semiconductor industry helped to educate many of us about the complexity of bilateral competition.

This project could not have been completed without the assistance of many individuals and organizations. Our deepest appreciation goes to those in both countries who kindly agreed to be interviewed, participate in our meetings, or comment on our draft chapters. We are particularly grateful to Warren Davis and Tom Hinkelman of the Semiconductor Industry Association, who gave generously of their time with unfailing good will, even though we sometimes disagreed on matters of interpretation or emphasis.

The project owes a special debt of gratitude to John G. Linvill, professor of electrical engineering at Stanford University, and to Takuo Sugano, professor of electronics engineering at the University of Tokyo, who served as co-chairmen. Their guidance and expertise were of crucial importance on numerous occasions during the course of the study. We could not have assembled this group of specialists, nor could we have won such extraordinary cooperation from both the U.S. and Japanese semiconductor industries without the active assistance of our distinguished co-chairmen.

We are grateful for financial support received from the U.S. Department of Commerce, the Henry Luce Foundation, the Japan–United States Friendship Commission, the U.S.-Japan Foundation, and, on the Japanese side, the Institute for Economic Research of the Japan Society for the Promotion of Machine Industry. We wish to emphasize, however, that responsibility for the contents of this report lies solely with the authors.

Finally, I wish to express our appreciation to Stanford's Northeast Asia–United States Forum on International Policy. The Forum, of which the Project on U.S.-Japan Relations became a part in October 1980, contributed both financial and administrative support. The latter came primarily from Nancy Okimoto, the Forum's Research Coordinator. Her hard work and patience, especially after my departure from Stanford in September 1981, contributed significantly to the completion of this book. I also wish to thank John W. Lewis, the Forum's co-director; Henry K. Hayase, administrative director of the U.S.-Japan Relations Program; and Anne Blenman-Hare, Helen Morales, Marilyn Webber, and Barbara Johnson, staff assistants.

F.B.W.

Contents

ONE   Introduction   1

TWO   Background   9

THREE   Technological Resources   35

FOUR   Political Context   78

FIVE   Finance   134

SIX   Conclusions   177

Appendixes   239

Glossary of Abbreviations and Technical Terms   249

Notes   255

Index   267

Tables

1. Degrees Granted Annually in Electrical and Electronic Engineering
in the U.S. and Japan, 1969-1980   30
2. Employee Turnover Rates in the U.S. Electronics Industry in 1979   61
3. U.S. Semiconductor R&D Expenditures, 1955-1961   81
4. U.S. Production (or Shipments) of Semiconductors for Defense
Consumption, 1955-1977   84
5. Government Procurement of Integrated Circuits Production,
1962-1968   85
6. Japanese Production of Integrated Circuits, 1978-1981   91
7. Japanese Government Subsidies for National Research Projects,
1966-1983   103
8. A Typology of Industrial Policy   124
9. Funding Sources for Japanese Industries, 1952-1975 and 1980   131
10. Indirect Financing in the U.S. and Japan, 1966-1978   141
11. Sources of Finance for U.S.-Based Semiconductor Firms,
1970-1979   142
12. Sources of Finance for Japan-Based Semiconductor Firms,
1970-1979   143
13. Adjusted Sources of Finance for Japan-Based Semiconductor
Firms, 1970-1979   145
14. Largest Shareholders of NEC, March 1980   147
15. Largest Lenders to NEC, March 1980   149
16. Estimated Capital Expenditures of U.S-Based Merchant Manufacturers (Semiconductors Only), 1973-1979   166
17. Estimated Capital Expenditures of Japan-Based Manufacturers
(Semiconductors Only), 1973-1980   166
18. U.S. and Japanese Integrated Circuit Production, 1978-1981   181

Figures

1. Contributions of U.S., Japan, and West Germany to International
   Solid State Circuits Conferences, 1971-1981   27
2. Storage Capacity and Demand for MOS RAM, 1970-1986   86
3. The Lifespan of Japan's IC Equipment, 1981   87
4. Average Transistor Prices in the U.S., 1955-1959   88
5. Costs of Capital in Japan and the West in Relation to Savings and
   Demand   139
6. Relation Between Internal Semiconductor Sales, Semiconductor
   Sales, and Total Sales of Selected U.S. and Japanese Companies,
   1978-1980   160
7. Relation Between Semiconductor Sales and Total Sales of Selected
   U.S. and Japanese Companies, 1980   163
8. Level and Growth of Capital Expenditures of Selected U.S. and
   Japanese Companies, 1979-1980   168
9. Semiconductor Sales and Capital Expenditures of Selected U.S.
   and Japanese Companies in 1980   171
10. Determinants of Semiconductor Industry Competitiveness   178
11. Semiconductor End-User Applications in Japan (1980) and the
    U.S. (1978)   180

Contributors

M. *Therese Flaherty*, Assistant Professor, Harvard Business School, Harvard University

Hiroyuki Itami, Associate Professor, Department of Commerce, Hitotsubashi University, Tokyo

John G. Linvill, Professor, Department of Electrical Engineering; Director, Center for Integrated Systems, Stanford University

Daniel I. Okimoto, Assistant Professor, Department of Political Science; Co-Director, Northeast Asia–U.S. Forum on International Policy, Stanford University

Takuo Sugano, Professor, Department of Electronic Engineering, Faculty of Engineering, the University of Tokyo

Michiyuki Uenohara, formerly Director of Research, currently Senior Vice President, Nippon Electric Company, Ltd.

Franklin B. Weinstein, President, Asia Consulting Associates, Inc., San Jose, California; formerly Director, Project on U.S.-Japan Relations

Competitive Edge

ONE

Introduction

Daniel I. Okimoto, Takuo Sugano,
and Franklin B. Weinstein

Not so long ago, U.S. manufacturers dominated the world in such key industries as steel, automobiles, and consumer electronics. American companies set industry standards, pioneered new technology, and controlled the lion's share of world markets. Japanese competitors hardly seemed capable of mounting a serious challenge, and Japan's economy, even as late as 1960, lagged far behind. Although the economy had recovered from the rubble of wartime defeat and had entered its income-doubling phase, most Japanese viewed their country as small, backward, dependent, and fragile. This perception stemmed largely from Japan's dependence on overseas markets and raw materials, an early shortage of capital to meet business investment demand, heavy reliance on bank loans for corporate financing, and the need to purchase technology from abroad. The whole economic edifice seemed like a house of bamboo, ready to collapse under the impact of the first serious crisis. Hardly anyone in Japan, and certainly no one in the United States, anticipated Japan's emergence as the economic giant it is today.

During the 1970's, Japanese companies succeeded in dislodging U.S. leaders in steel, automobiles, and consumer electronics. Nippon Steel, Toyota and Nissan, and Matsushita have apparently supplanted U.S. Steel, Ford and Chrysler, Sylvania and Zenith as world industrial leaders. So rapidly have the Japanese captured world markets, including those in the United States, that Western nations have had to resort to trade restrictive measures to protect domestic producers from bankruptcy and the danger of complete Japanese take-over. Parlaying heavy capital investments into modern, highly efficient plant facilities, the Japanese have managed to expand their world market share in one industry after another by offering top-quality merchandise at very low prices, with the promise of conscientious after-service. Japanese competitive strengths are so formidable, indeed, that they appear capable of dominating almost any sector they set their sights on.

What about high technology? Now that the Japanese have decided on shifting the locus of economic production from old-line manufacturing to the knowledge-intensive industries, will America's experience in steel, automobiles, and consumer electronics be repeated? Nearly all high technology industries of the postwar era—atomic energy, jet aircraft, artificial intelligence, computers, semiconductors—have sprung from America's military-related research and development system. High technology is one of America's last bastions of industrial supremacy. Will it, too, fall before the Japanese competition?

For years prior to the mid-seventies, this bastion had been considered unassailable. Even when the Japanese began showing remarkable recovery, Americans could still take comfort in the apparent lack of technological creativity in the Japanese. The alacrity with which the Japanese copied or purchased advanced technology from the West gave them a reputation as imitators, or at best, incrementalists in new product design and production technology. Of 500 technological innovations deemed "significant" between 1953 and 1973, the Japanese could claim credit for developing only 34, or slightly more than 5 percent, compared with 17 percent for Great Britain and 63 percent for the United States.[1] For the world's second largest market economy, the figure is strikingly low. It gives rise to questions as to whether the Japanese can compete when the pace of technological change is much brisker than it is in old-line manufacturing sectors. Are the Japanese capable of innovating at the frontiers of state-of-the-art technology? To what extent can the distinctive strengths of Japan's industrial system be applied to the high technology enterprise? Are Americans capable of responding effectively to the Japanese challenge?

The first test of bilateral competition in high technology is already well under way. It is the U.S.-Japanese struggle over semiconductors, especially the integrated circuits (ICs) on which thousands of individual transistors, diodes, and resistors are densely arrayed on a single piece of material (usually silicon). These tiny microchips perform the vital functions of arithmetic calculations and information storage and retrieval, which constitute the "brains" for computers, hand-held calculators, digital watches, robots, sophisticated machine tools, and other industrial and consumer products. As implied by its wide range of product applications, semiconductor technology is absolutely essential to competitiveness across a spectrum of industries: telecommunications, computers, machine tools, avionics, consumer electronics, and robotics, to name a few. (See Appendix A.) Figuratively speaking, semiconductors are the "crude oil" of the high technology sector, the indispensable core of advanced technology both now and in the future.

In 1982, world semiconductor production came to only $17.6 billion, but semiconductor components and technology played a key role in the $200 billion electronics industry. By 1990, semiconductor production is expected to reach upward of $40-60 billion, within an $800 billion-$1 trillion electronics sector. Moreover, semiconductor technology is an essential part of U.S. national security. It provides the technological basis for America's qualitative lead in nuclear missiles, precision-guided munitions, cruise missiles, surveillance and early warning systems, communications, aircraft, and an assortment of conventional weaponry, deployed to offset the USSR's quantitative edge in firepower. The stakes in semiconductors are therefore very high.

Although Japan's overall share of America's semiconductor market is still small (less than 7 percent), its surge in the U.S. 64K RAM merchant market (roughly 70 percent in 1982, not including in-house production by IBM, AT&T, and other captive suppliers) has alarmed many Americans. Some analysts believe that Japan's strong position in the 64K RAM merchant market will lead to domination of the 256K and subsequent RAM markets. Since mass volume memory chips offer a technological, production, and commercial base for advancement in other products, Japan's successes here have been taken as an inauspicious sign. Are the Japanese poised to capture bigger and bigger shares of the U.S. semiconductor market, one of the last citadels of American supremacy?

As the reader of this book will discover, Japanese domination of the semiconductor industry is by no means inevitable. Sounding the sirens now, as many in journalism and industry have done, is premature—though doing so may serve political purposes such as raising public awareness. The competition in semiconductors differs from that in steel and automobiles in the speed, importance, and complexity of technological change, more pronounced learning curve effects, and U.S. industry's greater awareness of the Japanese "threat." The rapidity of technological change makes it entirely possible for one manufacturer to be leading in a given area, say RAMs, while another is ahead in a different product line, like microprocessors, and for the situation to shift significantly in a relatively short time. The Japanese bring some impressive assets to this competition, but so do the Americans.

In this book, we analyze the comparative strengths and weaknesses of each country's semiconductor industry in terms of technology, political systems, industrial policy, capital availability, and financial institutions. The competition is not simply that of individual companies, and one cannot merely project outcomes on the basis of comparisons between, say, IBM and Nippon Electric Company. These companies operate within larger political-economic systems that have a central bearing on their competitiveness. Hence, the semiconductor showdown ought to be understood, in broad perspective, as national systems in competition. We focus on technology, government, and finances with one overriding question in mind: What impact do these factors have on the ability of each industry to compete successfully in the international marketplace?

The next two chapters take a look at the state of technology and national capabilities on both sides. In Chapter 2, Michiyuki Uenohara, Takuo Sugano, John Linvill, and Franklin Weinstein trace the development of the two semiconductor industries, exploring, among other things, their divergent approaches to R&D. The reader will no doubt notice substantial discrepancies in American and Japanese interpretations of this history. The legacy of divergent pathways of development is still discernible in areas such as industrial structure and technological orientation. The two semiconductor industries conduct R&D within two very different systems: the U.S. is dominated by military needs and the National Science Foundation's stress on fundamental knowledge, while Japan's system emphasizes commercial applications and economic advancement. Other chapters discuss some of the implications of this striking contrast between America's "military-industrial complex" and Japan's strictly commercial "bureaucracy-business-banking infrastructure."

In Chapter 3, Weinstein, Uenohara, and Linvill assess the relative positions of the two semiconductor industries so far as semiconductor technology is concerned. Their investigation confirms some often-heard generalizations about comparative technological advantage, but also shows the inaccuracy of some widely held assumptions, particularly about the Japanese. The authors show that in many key areas of semiconductor technology, American companies still hold a clear lead. American strength continues to lie in sophisticated new product designs like microprocessors, which the U.S. companies develop and commercialize before anyone else. Americans are also ahead in software, an area that accounts for a growing portion of value-added and is often decisive for the functioning of end-product systems like microcomputers.

The Japanese, however, excel at process and production technology. Chapter 3 describes efforts being made to build on Japan's already considerable base in manufacturing technology, and it also points out several potentially important areas where the Japanese have apparently gained an edge, belying widespread notions about their lack of innovative talent. In certain areas like software, for example, the Japanese may not be as far behind as many believe. Indeed, their apparent sluggishness probably has less to do with cultural impediments than with their late start. It may take years before the Japanese build up the huge inventory of software programs that exists already in America; but if the authors of Chapter 3 are correct, there is no intrinsic reason why, with the passage of time, the current gap cannot be closed.[2]

For years following the end of the Second World War, the Japanese were notorious for producing goods of poor quality. Thanks to the adoption of quality control techniques formulated by such American pioneers as William Deming and J. M. Duran, the problem was finally overcome, and today the label "Made in Japan" stands for quality workmanship and unsurpassed reliability. Hewlett-Packard once reported that 16K RAMs imported from Japan clearly outperformed American-made 16K RAMs in terms of reliability rates.[3] Japanese companies are said to have an easier time substituting capital for labor, partly because their system of lifetime employment lowers labor union resistance to job-displacing mechanization,[4] and partly because capital is more abundantly available for investment in new automated equipment. Are American-made integrated circuits inferior to Japanese circuits in quality? Or is this another false notion, like the myth of Japanese technological imitation? The authors of Chapter 3 discuss this question in some detail.

Competitiveness in the semiconductor industry is also affected by political institutions and public policies. Here again, Japan is said to hold a substantial—even "unfair"—advantage because its government and private sector are perceived to be a monolithic entity that works in collusion against foreign competition. Indeed, one reason the Japanese threat appears so menacing is precisely because Japan's "national conglomerate" purportedly confers an assortment of advantages on its semiconductor industry that America's decentralized, market-oriented government could never hope to duplicate: preferential lending, nationally organized research projects, flexible application of antitrust, and a "targeted" industrial policy that assigns priority to high-growth industries like semiconductors and provides a clear framework for the management of Japan's industrial economy. Some American executives feel, accordingly, that what the U.S. semiconductor industry is up against is not simply market competition from rival Japanese firms but nothing less than a national Goliath called "Japan, Inc.," against which American companies can hardly compete on equal terms.

How accurate is this portrayal of government-business relations in Japan? In what ways and to what extent has the Japanese government fostered the development of the semiconductor industry? How does this compare with the role played by the American government? These are a few of the questions that Daniel Okimoto addresses in Chapter 4. Looking at the historical development of the semiconductor industry in both countries, Okimoto identifies the types of policy instruments that have been effective at successive stages. He finds that the semiconductor industry, as a rapidly growing and technologically changing enterprise, tends to flourish in countries like the United States and Japan in which competition is vigorous and market dynamics drive commercial development. This suggests that the scope for effective government intervention is a good deal less broad than the idea of "Japan, Inc." would imply. If the government intervenes too heavy-handedly, market incentives can be distorted and private sector dynamism stifled.

The semiconductor industry as a case study of the high technology sector offers an intriguing glimpse into the efficacy of industrial policy. Japan's successful use of industrial policy has generated in America serious and high-level deliberations concerning the possible adoption of something similar in order to lift sagging industrial productivity and correct the inefficiencies currently caused by the protection and subsidization of declining, old-line industries. What elements of industrial policy have been used for the semiconductor industry? With what effects? To what extent, if at all, has U.S. industry been hampered by its government's disavowal of an official industrial policy? Would a Japanese-style industrial policy be compatible with America's political-economic institutions and values? Other subjects explored in Chapter 4 include assessments of comparative tax burdens, antitrust policies, government R&D, and the political context of economic policymaking. The picture here, as elsewhere, is complex; but as Okimoto argues, the Japanese government has probably promoted semiconductor development more effectively than its U.S. counterpart.

Competitiveness in the semiconductor industry is also linked increasingly to the capacity of companies to obtain capital for investment in R&D and new plants and equipment. Owing to rapid equipment obsolescence, fast technological change, and stepped-up competition, capital requirements have risen steeply since the mid-seventies. Companies that can raise capital more easily and at lower cost therefore have obvious advantages. In Chapter 5, M. Therese Flaherty and Hiroyuki Itami analyze corporate financing for American and Japanese producers of semiconductors within the context of two different financial systems. They address such key questions as: Do company characteristics, particularly size, diversification, and vertical integration, affect the availability, amount, and costs of capital? How do Japanese and American companies go about financing for fast-paced growth? Are there significant differences in capital market institutions? What are some of the implications of national differences?

Chapter 5 helps to clarify the question of why Japanese banks are willing to extend loans to highly levered Japanese companies when, under similar circumstances, American banks would consider the risks too great. It is not, as some U.S. business executives assume, because the Bank of Japan and the Ministry of Finance stand behind such loans (though loans from government banks are sometimes taken as implicit guarantees for those projects). Japanese banks are permitted to hold limited equity shares in companies to which they lend, unlike American commercial banks, which are prohibited from doing so by the Glass-Steagall Act (1933). Chapter 5 suggests that because of equity ownership, bank loans can be considered a form of "preferred stock," which implies a different definition and calculation of risk. The relationship between banks and corporations is, therefore, a much closer one than is customary in the United States—so close, indeed, that it can be described as a "banking-industrial complex" in its long-term and mutually interdependent nature.

Technology, government, and finances, in sum, are the important variables in what is clearly a complex and interrelated competition. The showdown in high technology is further complicated, as Okimoto points out in Chapter 6, by the proliferating network of transnational connections between American and Japanese semiconductor companies, including cross-licensing agreements, joint ventures, and direct foreign investments. The more that links of this sort are forged, the more blurred the national boundaries become. Ties of interdependence are, of course, important from the standpoint of managing the U.S.-Japanese alliance, especially as protectionist pressures intensify. But as long as there is competition, conflicts are bound to occur no matter how dense the cross-cutting linkages. Indeed, a striking feature of the semiconductor competition is the coexistence of conflict and interdependence; despite strong incentives for commercial cooperation, there are equally strong pressures to treat the bilateral competition as a zero-sum proposition.

Both sides see the stakes as high. Much is riding on the outcome— competitiveness in the high technology sector, future growth rates, employment spillover effects, military deterrence, alliance cohesion, and even, to some extent, the GATT-based system of free trade. Some American industrialists and government officials view the semiconductor competition in starkly zero-sum terms, even though the actual situation is more complicated. One executive has even gone so far as to remark: "If America loses to the Japanese in semiconductors, the consequences will be even more devastating than previous setbacks in steel and automobiles. There will be nothing left of comparative advantage except maybe farming."[5] This is an overstatement, of course, but it offers insight into why the U.S.-Japanese competition in semiconductors has attracted so much attention and has caused such great concern.

TWO

Background

Michiyuki Uenohara, Takuo Sugano, John G. Linvill,
and Franklin B. Weinstein

American and Japanese approaches to the development of semiconductor technologies have differed significantly. Before assessing the
impact of relative technological capabilities on the competitive position of the U.S. and Japanese semiconductor industries, we shall examine the contrasting approaches to technological development as the
two industries have evolved over the last several decades. The analysis
will focus on two issues: (1) the capacity for innovation, which em-
braces not only the design of new products and the development of
new product applications but also "process innovation" aimed at improving the efficiency and quality of production; and (2) the ability
of each country to maintain an adequate supply of highly skilled
professionals.

Semiconductor Industry Development in the United States

Beginning with the discovery of the transistor in 1947, the fundamental understanding of semiconductors was developed through a series of inventions at the Bell Laboratories. Although an enormous
amount of work has been done in this field around the world, no other
center has rivaled Bell Laboratories' contributions to the core technologies of modern semiconductor devices. From the massive telephone
industry of the United States, the semiconductor industry moved in approximately a decade to several large companies and then to a number
of new companies. Of the ten companies that were dominant in the
production of electron tubes in 1950, only two were still among the
ten largest producers of semiconductor devices by 1970. Entrepreneurial activity by gifted inventors and engineers was the key factor in
building the U.S. semiconductor industry. Even in some cases where
initial breakthroughs were made by engineers while they were still

working with larger companies, the knowledge that an innovative engineer could leave to start his own company was an important part of the incentive structure that encouraged innovation.

Semiconductor devices brought about very rapid increases in computing capability with increased volume and lower cost. Almost from the beginning, applications in the computer and digital systems area offered the principal market for semiconductor products. Moreover, in the telecommunications industry, a shift from linear to digital semiconductor devices occurred early.

A series of important developments in the U.S. rapidly extended the range of potential applications for semiconductors. The invention of the planar process by Jean Hoenni ushered in a new technology of photolithography and optical methods in the production process. In 1959, Jack Kilby, working in Texas, received a patent for his invention of the basic integrated circuit (IC) structure, and in the same year Robert Noyce, in California, invented a scheme for interconnecting circuit elements that made the IC practical.

A sequence of doublings per year of the number of active elements on a single integrated circuit began in 1960 and grew exponentially for almost twenty years at the same slope. At a pace unparalleled in any other technology, milestones in semiconductor technology followed one upon another. The invention of the microprocessor (a computer on a chip) by Ted Hoff at Intel in 1971 ushered in a series of very small and powerful computers whose evolution is still proceeding. The development of electron beam lithography in semiconductor processing, pioneered at the Bell Laboratories, IBM, Texas Instruments, and Hughes (and at Japanese and European companies as well), also opened up a broad range of new opportunities.

The integrated circuit was invented at precisely the time when an American response to Sputnik required new technology, and the needs of U.S. space and defense programs provided a unique opportunity for the development of these devices. The Minuteman missile program was committed in the early 1960's to the use of ICs, and this provided a base for the development of ICs at high reliability levels and in large quantities. The development of the semiconductor industry was significantly accelerated by the Minuteman program.

Large-scale U.S. government support for the semiconductor industry has by and large focused on a specific end product—for example, a missile system or a communications system. There has never been a broad program to develop technological potential for its own sake. Moreover, the Department of Defense, which in the 1960's was an important customer and in that way a stimulus to the semiconductor industry, has come to play a lesser role. It now uses only about 7 percent of semiconductor production, and that percentage is continuously decreasing. Also, as integrated circuits for military applications have tended to become highly specialized, the government has found it hard to persuade semiconductor manufacturers to address its specific requirements, as compared with demands originating from the very large industrial markets that have developed.

A more recent development is the Department of Defense's Very High Speed Integrated Circuits program (VHSIC), which plans to spend an estimated $300 million (during the first half of the 1980's) to develop new defense-related technologies. This program was set up to do research in the areas of advanced lithography, fabrication technology devices, system demonstration, and design/software/testing/architecture concepts. Specific military applications included radar, weapons control systems, code breaking, and enhanced resistance to radiation.

In early 1983, the Defense Advanced Research Projects Agency (DARPA) asked Congress for $50 million to launch a five-year, $500 million project to develop a "supercomputer," which, according to one senior U.S. Defense Department official, would outperform the "fifth-generation computer" that the Japanese were working to produce. Some reports indicated that expenditures on this project could run as high as $250 million a year. The program would target three areas of major interest to the computer field: materials technology, to devise faster and less power hungry circuits; computer architecture, to put those circuits together more efficiently; and implementation of advances in "artificial intelligence," allowing computers to go beyond "number crunching" to make human-style inferences.[1]

Industrial response to the VHSIC program and the DARPA supercomputer project was mixed. Supporters saw the program not merely as a source of profitable contracts but as a means of boosting U.S. R&D efforts in the VLSI (Very Large Scale Integration) area, and, in part, responding to government-sponsored programs in Japan and Europe. The VHSIC program promised to advance VLSI research down to the half-micron level; it also incorporated built-in test features, which are of particular importance in view of the extremely high costs of testing. Critics of the program, notably Robert Noyce of Intel, contended that the commercial spillover of VHSIC research would be slight. And because the industry has a serious shortage of skilled manpower, they predicted that the VHSIC program might simply have the effect of drawing scarce personnel away from projects of greater commercial relevance. The same criticisms have been voiced with regard to the DARPA supercomputer program. The initiator of the DARPA program acknowledged that the project would concentrate its attention on meeting military needs, not on spin-offs or competing with Japan. One leading U.S. executive saw "almost no product spin-offs" beyond certain advances in semiconductor processing. A number of observers questioned the ability of DARPA to spend such large sums of money wisely.[2]

Besides the development of new collective research efforts involving government support of industry, there has been growing industry support of university research, which has in the past depended almost exclusively on government funding. The Economic Recovery Tax Act of 1981 provided that 65 percent of corporate contributions to universities for basic research could be included among the expenditures qualifying for a general 25 percent tax credit for R&D expenses exceeding the average yearly amount of such expenditures during the three preceding tax years. Industry spokesmen have asked for a more generous tax credit that would cover the full amount of contributions to university research. Major centers for research on integrated systems have been established with industry support at such universities as Minnesota, Cornell, California, and Stanford. In such costly areas as advanced lithography and laser annealing, the support of university research, and the coordination of that research, could help reduce duplication of effort on the part of the companies.

In December 1981, the Semiconductor Industry Association (SIA) announced plans to establish a Semiconductor Research Cooperative (SRC), with each member of the SIA asked to contribute funds equal to one-tenth of one percent of its semiconductor sales. The goal was a fund of at least $10 million a year to support generic research on matters of long-term (three-to-ten-year payoff) interest, such as the use of electron-beam and X-ray technologies to etch integrated circuits. Current or near-future commercial projects were ruled out; the objective, according to SIA chairman Robert Noyce, was "to support areas of research we feel guilty for not doing." By spreading the funds among universities, the program would improve electrical engineering programs and help to increase the number of trained engineers. As Noyce emphasizes, "The SRC has as its main goal the production of brain power. Research results are secondary."[3]

Although few would question the desirability of encouraging industry support of university research on integrated systems, some critics of the program argued that academic efforts, which tend to be theoretical rather than applied, would be of very limited relevance to the commercial needs of industry. To some extent the criticism is justified: efforts undertaken by university centers will no doubt emphasize research more than development, for example by devising a test structure for electron beam lithography but not developing a complete system. But the results can still be of significant benefit to industry. Furthermore, under some of the programs that have been proposed, companies would be asked to supply members of joint industry-university research teams. For example, if a contract for computer-aided design were given to a university center, research fellows from the company could make up half of the research team. The companies might also sit on the steering committees that establish priorities for the university research centers. In some cases, a joint project might even be sited in one of the companies. Broader company involvement in this process would make it more likely that the research undertaken would have greater commercial relevance.

By mid-1983, the SRC had come to play a significant role in coordinating semiconductor research. The SRC, a non-profit organization, included most of the leading semiconductor and computer manufacturers. According to a senior SRC official, the cooperative was, in May 1983, signing or negotiating about 43 research contracts with 26 universities. The SRC expected to spend $11 million on research in 1983, $15 million in 1984, and over $30 million by 1986. Three types of research had already received funding by mid 1983: (1) permanent "centers of excellence" at several universities, with typical grants of $1 million annually; (2) project awards, each worth about $100,000, to selected professors for research in techniques for designing, producing, testing, and packaging integrated circuits; and (3) "new thrust" programs—for example, new chip fabrication techniques at the new Microelectronics Center of North Carolina, and three-dimensional chip structures at M.I.T.[4]

A second organization to foster coordinated research is the Microelectronics and Computer Technology Corporation (MCC), set up in January 1983. Backed by twelve major electronic and computer companies, the MCC was expected to have an annual budget of up to $100 million and to employ up to 400 scientists, engineers, and support workers. The MCC, a for-profit corporation with headquarters in Austin, Texas, would perform research in developing advanced computer architecture, software, component packaging, and computer-aided design/computer-aided manufacturing technologies. The member companies help pay the research costs and then share the research results.

Unlike the SRC, the MCC supports research done in the members' facilities. After the companies have shared the cost of developing new, high-risk technologies, they are expected to compete intensely to develop projects of their own. There have been efforts to formalize a relationship between the SRC and the MCC. One proposal was for a requirement that membership in the MCC be limited to companies that had already joined the SRC. Another proposal called for the SRC to function as the MCC's research arm in semiconductors.[5]

Semiconductor Industry Development in Japan

Japanese and American interpretations of the development of the Japanese semiconductor industry differ markedly. Even in this chapter, though the four authors agree that Japanese contributions to basic research and the development of new semiconductor technologies have been rather small, the Japanese coauthors believe that Japan has made significant contributions to the development of new manufacturing technologies and markets for semiconductor products.

Transistors. In the early days of transistor development, U.S. electronics companies developed new consumer products as potential applications for the transistor, but since most of those companies were already selling similar products using electron tubes, only a few companies, like Texas Instruments, which was not marketing electron tube products, were ready to move aggressively into the transistor market. It was owing to Japanese efforts, led by Sony's introduction of transistor radios in 1955, that the transistor and products using it became popular. Without those Japanese initiatives, the Japanese coauthors believe, the use of transistors might have developed much more slowly.

American industry sources have a different view of those early years. They point out that the major Japanese companies entered the semiconductor industry early, in the 1950's, and therefore have a longer history than most of the leading U.S. suppliers. Nevertheless, the Japanese were relatively unsuccessful in developing any applications other than in-house use. The technology was moving rapidly, and Japanese technology lagged by at least two years. Specifically, U.S. executives note that the Japanese were slow to make the transition from germanium transistors, which they built under license from General Electric, RCA, and Bell, to batch-produced silicon transistors. As a result, the Japanese suffered a major setback in the early 1960's. The Japanese established major facilities for the production of bipolar ICs during the 1960's, but according to U.S. industry sources, they were unable to compete effectively with companies like Texas Instruments and National Semiconductor through the early 1970's. Indeed, some Japanese accused U.S. firms of "dumping" in Japan at that time.

MOS circuits. The Japanese coauthors believe that a major contribution was made by Japan at the start of the MOS (metal oxide–semiconductor) IC era. During the early days of MOS IC development, there was concern about instability caused by complex problems at the interface between silicon and silicon dioxide layers. Because of these problems, according to the Japanese, U.S. companies were reluctant to invest heavily in production. Japanese semiconductor manufacturers, especially NEC and Hitachi, believed that new manufacturing technologies for mass production would assure product quality and reliability. Even though several American companies—including the Bell Laboratories, RCA, IBM, and Fairchild—played pioneering roles in the solution of these instability problems, the Japanese coauthors assert that Japanese companies were ready sooner than the Americans to undertake large-scale commercial production of MOS ICs. NEC foresaw a potential mass market for the desk-top calculator, and, in cooperation with Hayakawa (the predecessor of Sharp), developed calculators using MOS ICs. They completed a commercial model in 1966, and the success of this venture helped to establish the practicality of the MOS IC.

The boldness of the Japanese in establishing a large-scale production line when the commercial soundness of the MOS IC was still questioned by many people in the U.S. industry hastened progress toward the LSI (Large Scale Integration) and VLSI (Very Large Scale Integration) eras. On the other hand, the preoccupation of the Japanese companies with production of MOS ICs for desk calculators made them unresponsive to other opportunities. For example, Intel's pioneering work in microprocessor development was stimulated by a problem posed by Busicom of Japan, because Busicom was unable to get cooperation from Japanese companies that were preoccupied by the production of MOS ICs for desk calculators.

Thus, the Japanese coauthors conclude, although the Japanese semiconductor industry has not made any significant breakthroughs in the form of basic discoveries and inventions, it has contributed to realizing the broader potential of new semiconductor technologies. The R&D efforts of most Japanese semiconductor manufacturers have relied heavily on basic technologies developed in the United States, but Japanese semiconductor technology is not a mere copy of American technology. Most Japanese companies were putting considerable effort into the development of IC technologies even before many of the better known U.S. semiconductor companies were established. For example, NEC began IC development in 1960 and established P-channel MOS technologies in 1964; as already noted, NEC began manufacturing P-channel devices for desk-top calculators in 1966. The Japanese developed a family of ICs for electronic switching systems in 1967; a 144-bit N-channel MOS IC memory in 1968 (when most U.S. companies were developing P-channel MOS IC memories); and a 1-kilobit N-channel memory in 1971, which was a challenger to Intel's 1-kilobit P-channel MOS IC memory. Since then, the N-channel MOS IC has gained popularity. Japanese efforts in the area of linear IC families for consumer applications are well known.

The foregoing Japanese account of MOS development is disputed by U.S. industry sources. Early MOS devices, American executives say, were well suited to the Japanese market, with its emphasis on consumer products and calculators. Although calculators and watches using MOS circuits were first produced in the United States (Texas Instruments, Bomar, H. P. Frieden, and North American being the major producers), the Japanese saw the markets for these high-volume consumer products as fitting their product strategies well. Thus, they aggressively pursued those markets.

The typical pattern, as recalled by a U.S. executive, was for a Japanese company to team up with an American MOS producer to secure an initial source of MOS circuits. After bringing the product to market, the Japanese companies generally switched to a Japanese supplier of MOS circuits (either their own company or another) for later products.

United States industry sources emphatically deny that Japanese companies committed themselves to large-scale MOS production before American companies did so. In IBM and AT&T, the manufacture of MOS circuits occurred on a broad scale in late 1969. General Microelectronics (GME) was supplying MOS circuits to Frieden in the 1960's, and Hayakawa's initial source of MOS circuits was Rockwell. The Rockwell contract called for technology transfer to Hayakawa, and that was accomplished in due course. Rockwell, Texas Instruments, Mostek, American Microsystems (AMI), and others were major suppliers of MOS circuits to Japanese calculator manufacturers until Japanese suppliers became available.

One U.S. executive specifically questioned the Japanese claim that Busicom turned to Intel because Japanese companies were already fully committed to the production of MOS ICs for desk calculators. Busicom may indeed have been unable to gain the cooperation of Japanese MOS producers, given the extent of vertical integration in Japa-

nese companies, but Busicom's application was the desk-top calcula-
tor. If Japanese companies had been fully committed to producing
MOS ICs for desk calculators, Busicom should have been able to find a
Japanese source. It could not do so, however, and Busicom used
Mostek ICs for the hand-held calculator and the Intel microprocessor
(its first application) for the desk-top calculator.

When Japanese MOS suppliers saw a large local market developing,
they made major commitments to produce ICs for that market. It is
simply fallacious, asserts one American executive, to imply that the
sequence was reversed. The Japanese did not show boldness in undertaking mass production for a market that U.S. companies could not
foresee; rather, the Japanese companies moved to mass production of
MOS ICs only after such a market had already begun to emerge, and
U.S. producers had been instrumental in bringing that market into
being.

After the Japanese began to produce MOS circuits on a large scale
for desk-top calculators, the U.S. producers who had previously been
the principal suppliers of the Japanese gradually lost interest, for two
reasons. First, the American companies found they were losing their
domestic customers. Second, they found selling in Japan difficult in the
face of what they saw as a culturally engrained "Buy Japan" attitude.
Indeed, at that time (the early 1970's), the import of complex ICs into
Japan for distributor stock was not permitted; imports were only allowed for previously specified end customers. The Japanese Ministry
of International Trade and Industry (MITI) made this import data
available to Japanese suppliers, and in that way stimulated and directed the efforts of Japanese IC producers.

The real success of Japanese producers, American industry sources
conclude, came only after the mid-1970's. MITI targeted the computer
and telecommunication markets as central to Japan's future. Establishing a national goal to lead in those industries, the government offered
substantial incentives to encourage R&D and investment, besides restricting foreign access to Japanese markets.

Basic research. There is now a broad recognition among the Japanese of the urgent need to promote R&D activities aimed at increasing
Japan's contribution to innovation in basic technologies, and many
Japanese companies have substantially increased their expenditures
for basic research. These efforts are unlikely to match basic research
outlays by the larger U.S. companies, but they are a beginning.

A major thrust of Japan's effort to contribute to basic semiconductor technology has been the well-publicized program for industry-
government cooperation in VLSI research. Actually, there have been

two VLSI products, one directed at communications applications, and the other at computer applications. The communications-oriented project began in 1975 under the management of the Electrical Communication Research Laboratories of Nippon Telegraph and Telephone Public Corporation (NTT), which falls under the administrative guidance of the Ministry of Post and Telecommunication. NEC, Hitachi, and Fujitsu cooperated in this project, and government support was provided in the form of procurement. NTT did not give direct financial support, and the three cooperating companies invested their own R&D money on the supposition that the investment would be recovered through future procurement. In the first phase of the project, which ended in the 1977 fiscal year, the major objectives were to investigate the practical limit of photolithography and to study basic micron and submicron device technologies. The second phase, lasting for another three years, applied the results of the first phase to develop special purpose LSI for communications and to carry on the development of other new technologies for communications.

The more publicized computer-oriented project, partly founded by MITI, ran from 1976 to 1980. The objective of this project was to develop VLSI technology as a key to the future development of computer systems. As is well known, MITI has singled out computer development as an area of highest priority for the Japanese economy in the 1980's. The project budget was around 70 billion yen (about $200 million) over the four-year period. About 60 percent of the total budget was financed by the five member companies; the rest came from MITI in the form of interest-free loans to the member companies, to be repaid from royalty income and profits derived from the products that resulted from technologies developed by the project. (Some U.S. commentators have suggested, however, that the Japanese companies may be able to avoid having to repay these loans.)

Unlike NTT, MITI is not in a position to offer procurement as an inducement to companies. MITI's objective is to promote industrial R&D, partially filling the gap caused by the absence of any significant amount of defense-related R&D in Japan. Thus the management of this project was left largely to industry, which established a VLSI Technology Research Association for that purpose. Associated companies included NEC, Toshiba, Hitachi, Fujitsu, and Mitsubishi. The research was carried out at three laboratories: the Association's cooperative laboratories; the Computer Development Laboratories (CDL) established jointly by Hitachi, Fujitsu, and Mitsubishi; and NEC-Toshiba Information Systems (NTIS) Laboratories owned jointly by NEC and Toshiba. The cooperative laboratories were staffed by research engineers from the five companies and from the Electrotechnical Laboratory of MITI's Industrial Science and Technology Agency. The CDL and NTIS laboratories drew their personnel from the companies associated with each of those laboratories. Only the cooperative laboratories actually existed at a single site. The CDL and NTIS laboratories were scattered on the premises of the member companies.

Research was concentrated on microfabrication, crystal, design, process, test and evaluation, and device technologies. The cooperative laboratories were responsible for common basic technologies, and the CDL and NTIS for application technologies. Although many people are under the impression that the cooperative laboratories carried the major portion of the Association's R&D activities and that the purpose of the project was to develop the final VLSI products for the future development of computers, the cooperative laboratories only carried on research that was related to common basic technologies, which represented a minor portion of the Association's overall activities. Since the project was set up to develop technologies from which the individual companies would later develop specific products, product development was left entirely to the companies involved and has had to be undertaken with company funds. Even with strong administrative guidance from MITI, the member companies, which are all intensely competitive with one another, could not agree on final design and processes. It is widely believed, furthermore, that this intense competition made the participating firms reluctant to send their best engineers to the cooperative laboratories.[6]

In the various cooperating laboratories, each research team included engineers from all member companies, but in the CDL and NTIS all development teams except those concerned with standard basic technologies consisted almost exclusively of engineers from a single company. Device technology in particular is very specialized within each separate company, and mixing engineers from different companies would have been unworkable, to say the least. Except for the cooperative research laboratories, it was agreed that there would be virtually no technical exchange through reciprocal visits between the CDL and NTIS; even among the companies associated with one or the other of those two group laboratories, such visits were largely ruled out. The exchange of information took place through written reports, technical committee meetings, and periodic project symposiums.

After the conclusion of the research project at the end of March 1980, the participating companies continued working on their own to improve, modify, and commercialize the basic technologies developed through the project. Since the project intentionally avoided any effort to develop specific products, it was more successful than initially anticipated. At the outset, members tended to be very suspicious of one another, and communication was extremely poor, but that changed dramatically as the result of extensive discussions, which led to agreement on objectives and detailed plans. And since technologies that required interfacing with traditional technologies of a particular company were kept out of the joint projects and were developed instead by individual companies, no company had to risk revealing particular technologies. Only basic technologies and information were freely exchanged.

Since VLSI technology is very expensive, the joint development of various alternative technologies in order to identify the most promising ones for further development proved to be a highly efficient use of resources. A comparable result could have been achieved in universities, if they had had the funds and were willing to undertake such research and development. In Japan, where university involvement in basic research on material physics and process technology is limited and university-industry cooperation is generally poor, that alternative was virtually out of the question. Thus, industry had to devise a new way of doing research and development in such resource-intensive areas. In the Japanese view, the VLSI project demonstrated that such cooperative research efforts could be carried out effectively.

As a result of the project, more than 1,000 patents were issued, and about 460 technical papers were published, many of them as part of international conferences in the United States, Europe, and Japan. These results fell into five major technical areas: design methodology, electron beam lithography, silicon crystal growth and processing, device testing, and VLSI devices. Clearly, the VLSI project was a remarkable organizational innovation that essentially succeeded in engendering a new breadth of capability in the Japanese semiconductor industry.

The VLSI project is not the only instance of industry-government cooperation to advance technological development in the Japanese semiconductor industry. MITI also organized a project, with anticipated government expenditures of 23 billion yen (approximately $100 million) over the 1979-83 period, to upgrade Japanese capabilities in basic software technology and peripheral equipment. By May 1983, MITI's software development program, said to be only in its third year, had received $192 million in funding, half from the government.[7]

In mid-1981 the Japanese government announced the creation of a new research and development institute to develop basic technologies for optoelectronic ICs (OEICs) during a six-year period from 1981 to 1986. This joint industry-government R&D project, in which engineers and scientists from nine leading Japanese companies work with counterparts from the government's Electrotechnical Laboratory of the Industrial Science and Technology Agency, concentrates on optically based industrial measurement and control systems. OEICs are regarded as holding the key to future realization of super-high-speed, highly efficient computers that utilize optical signals. OEICs include laser diodes, photo detectors, and logic circuits integrated on the same chip. One of the most important aspects of research in the project concerns ways of growing gallium arsenide crystals to replace silicon crystals as a substrate for OEICs. The government's financial contribution was estimated at $82 million over the six-year period.[8]

In addition to the OEIC effort, the Japanese government has organized and financed efforts to develop super-high-speed device technology, including development of the most advanced Josephson junction circuits.[9] According to one report in early 1983, major Japanese companies were working on a $104 million MITI project to develop a computer by 1989 that runs ten times faster than any machine under development in the United States.[10] A Japanese government-sponsored program to develop "fifth-generation" computers, with a prototype computer with artificial intelligence—the ability to reason—targeted for 1990, is reportedly expected to cost some $400 million over this decade.[11]

According to one U.S. academic expert on computer technologies, the Japanese were, by mid-1983, poised to make great strides in almost every area of computer technology. Development projects under way in Japan on artificial intelligence, computer-aided design and manufacturing, and large-scale numeric processors all reflected the long-term orientation of the Japanese.[12]

U.S. and Japanese Approaches Compared

A review of the history of semiconductor technology development in the United States and Japan shows a basic difference in approach between the two countries. In the Japanese industry, development was mainly a response to consumer market demand, whereas the main stimulus to the American industry's early development came from demand generated by high technology markets, including the computer industry and government procurement requirements for military and space programs. Of course, the U.S. companies also put considerable effort into the development of consumer electronics, but the predominant commitment of resources was aimed at the high technology market. Of course, the Japanese did make major efforts to develop high technology markets, especially in communications and computers, but most Japanese engineers were oriented toward consumer electronics.

Government procurement. Most Japanese commentators are inclined to place considerable weight on the role of military procurement in stimulating the growth of the U.S. semiconductor industry. They note, for example, that more than 95 percent of U.S. IC production in 1964 was supplied to the government. American observers, on the other hand, emphasize that although the military did provide an important early market for silicon transistors and ICs, the main stimulus for the industry's growth was the commercial computer market. Government procurement may have been very high in 1964, but that situation was short-lived, and it is misleading to view it as an indication of the overall importance of government procurement. In the United States, computers pioneered the development and application of semiconductors, and the computer industry played a major role in setting high standards of quality and performance for the semiconductor industry, besides stimulating the exploration of advanced processes, tools, and devices.

Government-supported R&D. There is no question that the pattern of R&D support in the two countries differs sharply. Up until 1970, government-supported R&D was an important stimulus to the expanding U.S. semiconductor industry. Since then the U.S. government's role has receded. In Japan, government support for semiconductor R&D has only been of importance since 1972. Prior to 1972, the percentage of government R&D support was less than 2 percent of the total; in 1973 it increased to about 15 percent, as a result of the decision to appropriate funds as compensation to the industry for opening up the Japanese IC market. These funds were appropriated for only two years—1.7 billion yen in 1973 and 1.8 billion yen in 1974. Government support peaked at about 26 percent in 1977, reflecting both the increase in funds for the VLSI project and the stagnation of industrial investment resulting from a slowdown of the economy. Even then, government support was predominantly aimed at promoting the Japanese computer industry, and the government provided very little, if any, money for the development of consumer electronics. Government support of R&D dropped to less than 2 percent after the end of the VLSI project in March 1980, but, as already noted, new projects on OEICs and fifth-generation computers are receiving significant government funding during the eighties.

In the United States, the Department of Defense was quick to recognize the potential importance of the transistor and invested large amounts of money in its development. In the year 1954 alone, Department of Defense R&D funding for semiconductor technology was greater than the total amount of funding given the Japanese industry by MITI up to 1974. Also, because R&D undertaken for U.S. military and space programs was not only risky but extraordinarily expensive, the government paid the full costs plus a fixed percentage profit. This cash inflow made it possible for semiconductor technology to move ahead quickly, and some of the achievements of defense-related R&D programs, such as the high precision photolithography technology developed for microwave printed circuits in antimissile radar systems, proved to have a broad impact on the development of IC technology.

Industry sources in the United States are inclined to play down the importance of government-sponsored R&D. They say that such support was never, even in the early years, as critical to the industry as many Japanese believe. Fairchild, for example, the IC leader in the early 1960's, did not participate in the Minuteman missile program; according to one executive who was at Fairchild in the early 1960's, government-sponsored R&D there never accounted for more than 4 percent of the total. Furthermore, most of the major advances in semiconductor technology, including the planar transistor, the IC, the dynamic random access memory (RAM), and the microprocessor, were achieved without any government support.

The Japanese coauthors believe that the U.S. approach had a major disadvantage—namely, that many companies with lucrative government contracts were overcautious about extending the new technologies into markets where the level of uncertainty was high. In other words, although the fact that many of the engineers who pioneered the IC industry's development left bigger firms with heavy government R&D contracts to establish new companies reflected the vitality of the American system, it also showed the conservatism of the large companies in assessing the potential for the merchant LSI market. U.S. executives contend that this interpretation simply is not in accord with the realities prevailing in U.S. companies at the time.

On the other hand, although there were, of course, no government R&D funds for defense and space technologies in the Japanese industry, Japanese companies that concentrated on stable markets, such as telecommunications technology produced for NTT, a public corporation, initially showed the same conservatism that the Japanese ascribe to the larger U.S. companies. Since they had to emphasize high performance, these Japanese companies were cautious in adapting transistors to lower-priced consumer products. But a second group of Japanese companies, excluded from NTT contracts, had been oriented toward consumer electronics from the start. As consumer markets began to develop, the NTT-oriented companies recognized the need to enter the consumer electronics market. Since the recovery of high technology development costs was so slow in Japan, those companies concluded that they had no choice but to diversify their product lines and take advantage of economies of scale. In general, it may be said that Japanese companies, obliged to rely on their own limited funds for R&D in the semiconductor industry, felt pressed to develop a wide range of markets in order to hasten the recovery of their investment.

Japanese and U.S. executives disagree on the importance of current government support for R&D. From a Japanese standpoint, even though U.S. Defense Department R&D contracts with semiconductor companies are greatly reduced from the levels of the 1960's, they are still significant in comparison with Japanese government contracts, which usually cover only direct expenses, without any fixed percentage of profit. Moreover, the U.S. Defense Department's VHSIC program will have pumped more government funds into semiconductor research than the ongoing or contemplated programs for industry-government cooperation relating directly to the Japanese semiconductor industry. In the case of the computer-oriented VLSI project, the Japanese government covered less than half of the total expenses. At the same time, however, because the technologies promoted by the Japanese government have been more directly applicable to commercial enterprise, government assistance has had a stronger impact on the competitive position of the Japanese industry—and it is this phenomenon that lies behind the rather common attitude among U.S. executives that the Japanese government gives much more effective financial support to its semiconductor industry than American companies receive from Washington.

Industrial structure. Besides differences in the nature of industry-government relationships and market orientations, the pattern of technological development in the U.S. and Japanese semiconductor industries may be said to reflect basic differences in industrial structure. In the United States, as already noted, most of the large diversified electronics companies that invested heavily in the development of semiconductor devices in the 1950's were slow to make a major commitment to ICs. They limited themselves to production of special-purpose ICs for in-house systems applications and relied on merchant IC producers for general-purpose IC components. During the years when capital requirements were relatively modest and the technological expertise for manufacturing a new product could be mustered by bringing together a few core engineers, small venture companies made significant contributions to the development of semiconductor technology. But as the industry began moving into the VLSI era, the dramatic increase in the amount of capital and technology needed to start a new venture made it more difficult for small companies to play a central role in the production of general-purpose components. The many small companies established with venture capital in the last several years have tended to focus on specialized applications.

In Japan, semiconductors are produced not by specialized semiconductor companies but by the semiconductor divisions of large electronic systems companies. These companies—mainly manufacturers of communications equipment, computers, and consumer electronics—have all been producing semiconductor devices since soon after the invention of the transistor. More recently, manufacturers of watches, electric components for automobiles, and desk-top calculators have become semiconductor producers.

Because of this pattern, the semiconductor producers not only have access to low-cost capital but also can take indirect advantage of tax incentives and other special benefits extended to other divisions of the company, and they are assured of selling at least a portion of their products to their own systems divisions. All this puts the Japanese companies in a generally better position than most of their U.S. counterparts—with the exception of IBM and Western Electric and one or two others—to raise the enormous amounts of capital required to pursue R&D in the VLSI area and to fund VLSI production facilities. But the smaller U.S. companies, often funded with venture capital and managed by engineers who are close to the technology, may be more willing to take risks to move in new technological directions and may also be more flexible in responding to narrowly defined, but commercially rewarding, opportunities to develop new applications of semiconductor technology.

Approaches to technology transfer. The contrasting patterns of technological development in the U.S. and Japanese semiconductor industries have also reflected differences in the two countries' approaches to technology transfer. Some of the impediments to the free flow of technology between the United States and Japan relate to differences in the way technology is transferred within each country. In the United States, technology transfer is based on the principle of the open market, with a free exchange of information and movement of personnel from one company to another. In Japan, owing to social constraints and lifetime employment practices, movement of personnel between companies is extremely rare. This makes it very difficult to assemble the sort of top-flight engineering team, drawn from a number of companies, needed to push back the technological frontiers and develop new products. Indeed, it was this inability to draw together engineers from various companies that led MITI to initiate the VLSI cooperative research project. But, as already noted, the work done in the cooperative research laboratory was only a part of the VLSI project's efforts; in some cases, participation in the project merely provided the member companies with information about new developments much earlier than nonmembers. This is an improvement, but it hardly compensates for the difficulties faced by the Japanese when they seek to draw together superior research teams from a number of companies.

For much of the period under discussion, formal academic conferences did not play the same role in promoting the effective exchange of information in Japan as in the United States. Japanese engineers tended to be awkward in presentation techniques, and the conference atmosphere was usually formal; debate was rare, and casual "lobby exchanges" were practically nonexistent. Recently, however, academic conferences in Japan have also come to play an important role in the exchange of information.

Up to now, the international flow of technology has been mainly from the United States to Japan. Since the United States is not only the center of modern science and technology but also in a convenient geographical location, most international conferences are held there. But technology is beginning to flow in the other direction as well, with the rapid increase in Japanese contributions to international conferences and American technical journals (see Fig. 1). Since these technology flows take place almost entirely in English, the Japanese must make a considerable effort to expand the transfer of information to the United States. While the Japanese have assiduously learned about American technology and industrial practices by their study of English and mastery of the capability to communicate effectively with Americans at all levels, few Americans have made a comparable effort to understand the Japanese language and Japanese ways of communication. If they did so, the exchange of technology could be expanded both ways to

Fig. 1. Contributions of U.S., Japan, and West Germany to International Solid State Circuits Conferences, 1971-1981.

mutual benefit. The dramatic growth in the number of visits by American engineers to Japanese research institutions is an encouraging sign, however.

Recently there has been some concern in Japan that the United States may tighten restrictions on access to military-related high technology. The Japanese generally agree that the only way they can avoid such restrictions is by developing comparable capabilities in high technology in order to facilitate fair exchanges. Commercial technology and patents have customarily been transferred by means of technical license agreements. Recent agreements have provided mainly for cross-licensing, which indicates that the technology flow is becoming more reciprocal. Although Japanese trade in technology is largely in deficit to the United States, most of that deficit represents payment for technologies imported many years ago. Now those payments have increased substantially owing to the expanding sales of products based on the use of those technologies.

Participation in the development of new IC technologies seems to be a growing source of friction between Japan and the United States, however. Some U.S. companies in Japan complained about being excluded from the MITI-sponsored VLSI project (though, as the Japanese pointed out, no U.S. company formally applied to MITI). The Japanese explained that the exclusion of U.S. companies was based not on national considerations but on the principle of selecting only a few companies. It is true, as the Japanese say, that a number of Japanese companies were denied participation, but an American executive observed that the only major Japanese company excluded from the VLSI program was Oki, and he suggested that this was because of Oki's joint venture with Sperry Rand.

The same problem has arisen in regard to the Japanese government's project to develop optoelectronic ICs. Should TI, Motorola, Intel, Fairchild, and other U.S. companies that either operate or plan to build manufacturing facilities in Japan be allowed to participate in joint industry-government programs to develop advanced technology? MITI officials responded that requests to do so could not be granted unless the United States made a reciprocal exchange of certain classified technologies such as electronics equipment and software for the P3C anti-submarine warfare aircraft. (The Japanese believe that the United States fears Japan will use such technologies for commercial purposes.) The MITI officials also asserted that the U.S. facilities in Japan do not have research and development departments and are therefore not qualified to participate in the development of advanced technology. According to U.S. executives, both IBM and TI have development programs in Japan, and RCA has a research effort. (Apparently what qualifies as R&D to the United States is not the same as what MITI calls R&D.) MITI did, however, make an offer to open the patents emerging from industry-government R&D projects to American concerns after the new processes are completely developed.[13]

While there are still impediments to the transfer of technology among Japanese companies and between Japan and the United States, it is likely that the future development of semiconductor technology in the United States will be more influenced by Japanese contributions than in the past. In part, this reflects Japan's enhanced technological capabilities. The current expansion of Japanese semiconductor plants in the United States, the growth of American design and production facilities in Japan, and the development of more cooperative relationships between U.S. and Japanese semiconductor manufacturers all increase the prospects for a more intensified, and more balanced, flow of technology between the two countries.

Training of New Professionals

The shortage of trained professionals may prove to be a significant constraint in the development of semiconductor technology in both Japan and the United States, but the problem is much more serious in the United States. In this section, we shall review trends in the production of skilled personnel for the two countries' industries, the nature of the training received in each, and some innovative approaches to training involving closer cooperation between industry and the universities.

Trends in the supply of engineers. The increasing sophistication of semiconductor technology has demanded an increasingly higher and broader sort of training for new professionals—individuals trained in both semiconductor and computer or communications technologies, and with advanced degrees. The working professional degree is now the master's degree, rather than the bachelor's degree, and there is a growing demand for engineers trained at the doctoral level. These professionals are in short supply.

The Sputnik challenge and the space programs of the 1960's provided a powerful stimulus to graduate education programs in semiconductor technology in the United States, but by the start of the 1970's the Vietnam war had induced an antitechnology backlash in U.S. universities. The result was a very substantial diminution in the number of engineering students in both graduate and undergraduate programs. During most of the 1970's the number of electrical engineering degrees granted in the United States declined; only in 1979 did the number of electrical engineering graduates approximate the number produced a decade earlier (see Table 1). At the same time, the number of electrical engineers trained in Japan rose steadily. Japan, with a population base approximately half that of the United States, was producing at the beginning of the 1970's the same number of electrical engineering bachelor's degrees as the United States; by the end of the decade, Japan was producing 50 percent more than the United States. It is clear that engineering has higher status in Japan than in the United States. For a period of time, American engineers referred to themselves as applied scientists; in Japan, the engineer has higher status than the scientist.

But even in Japan, there has been some concern about the supply of electrical engineers. Statistics on the number of students pursuing studies in electrical engineering (as opposed to degree recipients) indicate a slight decrease at both the undergraduate and graduate levels. This reflects a more general decrease in the number of university students because of declining birth rates.

According to a 1981 survey undertaken by a panel organized by the American Electronics Association (AEA), the shortage of electrical engineers in the United States was expected to become an even more serious problem as the industry expanded: companies anticipated the need for twice as many engineers to fulfill their plans for expansion.

TABLE I

Degrees Granted Annually in Electrical and Electronic Engineering in the U.S. and Japan, 1969-1980

Year	United States				Japan			
	B.S.	M.S.	Ph.D.	Total	B.S.	M.S.	Ph.D.	Total
1969	11,375	4,049	858	16,282	11,035	705	108	11,848
1970	11,921	4,150	873	16,944	13,085	688	116	13,889
1971	12,145	4,359	899	17,403	14,361	844	109	15,165
1972	12,430	4,352	850	17,632	16,020	913	119	17,052
1973	11,844	4,151	820	16,815	16,205	1,026	114	17,345
1974	11,347	3,702	700	15,749	16,140	1,173	106	17,419
1975	10,277	3,587	673	14,537	16,662	1,258	120	18,040
1976	9,954	3,782	644	14,480	16,943	1,201	114	18,258
1977	9,837	3,674	574	14,085	17,868	1,447	142	19,257
1978	10,702	3,475	524	14,701	18,308	1,686	132	20,126
1979	12,213	3,335	545	16,093	19,572	1,697	166	21,435
1980	13,745	3,740	523	18,008	NA	NA	NA	NA

SOURCES: Engineering Manpower Bulletin (U.S.), Ministry of Education (Japan).

The AEA estimated that there would be a shortage of 129,000 electrical and computer engineers by 1985. Although this estimate was challenged by the Institute of Electrical and Electronics Engineers (IEEE), there is no question that the problem is of some magnitude. Engineering schools can be expected to produce about a third more engineers than current levels, and the IEEE suggests also better use of older engineers; but more must be done.

The problem now is not that of encouraging undergraduates to study electrical engineering; engineering schools already have more applicants than they can accept. Rather, the problem is the lack of teachers. To some extent this shortage reflects the antitechnology backlash of the seventies, but there are additional causes. The universities, financially pressed, have tended to assume that the present upsurge in demand for electrical engineering is temporary, and for that reason they have been reluctant to increase the number of faculty positions in electrical engineering. Nor can they, in most cases, meet the salaries offered by industry.

The salary problem also affects enrollment at the Ph.D. level. As Table 1 shows, during the 1970's the United States produced many more Ph.D.s in engineering than Japan did. But the increasing salary scale of the semiconductor industry is drawing many holders of undergraduate degrees into industry. Obviously, if students stay on for graduate programs, they must sacrifice the potential income they might have earned during the time of their education; they have also been finding that the salary differential attached to a higher degree is less than it was a decade ago. This may be changing, as the demand for Ph.D.s increases. Also, there is a growing number of programs enabling employees of electronics companies to continue their education on a part-time basis during their first years of employment, usually with financial support from their companies. But this does not solve the problem of producing more Ph.D.s oriented toward teaching careers who might help alleviate the shortage of university faculty.

In response to this problem, the AEA has a plan, which was announced in 1981, to establish a national foundation to assist engineering education in the United States, starting with an initial contribution of $50,000 from its operating revenues and 2 percent of annual members' dues every year. The plan calls for additional contributions from member companies in an amount equal to 2 percent of their R&D budgets. This would amount to as much as $30 million to $50 million annually. The funds may be used to create engineering faculty chairs, to provide part-time and visiting faculty, and to support faculty fellowships. Contributions need not be in cash, but can be in people, facilities, and equipment donated to educational institutions.[14] In addition, the Exxon Foundation has announced $15 million in grants to entice engineering students into teaching careers.[15] It may well be that the most cost-effective approach would be to enlist industry personnel to serve as part-time or adjunct faculty, with the companies continuing to pay their employees' full salaries.

In Japan, universities have not had any serious difficulties in hiring good electrical engineering teachers. The starting salary is about the same as the starting salary in industry, and nearly all engineering faculty members, including lecturers and assistants, are hired with tenure. There has been a steady increase in the number of faculty at Japanese engineering schools. It should be noted that assistants at Japanese universities are mostly Ph.D. degree holders, who are assigned tasks roughly corresponding to those performed by research associates at U.S. universities.

U.S. and Japanese training compared. American universities place more emphasis on advanced graduate training in electrical engineering than do Japanese universities. Although the rate of production of advanced degree holders has declined in the United States over the last decade, it is still very much larger, even on a per capita basis, than in Japan. Moreover, the supply of computer engineers in the United States has grown continuously, and there has been an impressive increase in the number of students pursuing graduate study in computer engineering.

In Japan, computer engineering has developed as a component of electrical engineering programs. Thus, Japan in 1979 produced a total of 166 Ph.D.s in electrical engineering and computer engineering combined, while the United States produced 545 electrical engineering Ph.D.s and 190 Ph.D.s in computer engineering, for a total of 735. The Japanese produced a total of 1,697 master's degree holders, while the United States produced 3,335 master's degrees in electrical engineering and 1,974 in computer engineering, for a total of 5,309. By the early 1980's, however, the Japanese were rapidly closing the gap at the master's degree level. It is also worth noting that in Japan an engineer working in industry may submit a doctoral dissertation based on work done for his company; he is not required to be enrolled in a university while undertaking the research for his Ph.D. dissertation.

In 1981, eight major national universities in Japan had Ph.D. programs in electrical engineering, including information processing and computer engineering. Some offered undergraduate instruction as well, whereas the others admitted only graduate students. The situation is very different in the United States, where the major producers of Ph.D.s in electrical engineering among American universities almost uniformly have undergraduate programs. Practically every state university, and every major private university with a school of engineering, offers a Ph.D. in electrical engineering. Computer engineering programs in the United States are somewhat different, in that a number of departments of computer science have sprung from departments of mathematics, and some of these offer only graduate programs. But the largest source of computer engineering work is in conjunction with electrical engineering departments, and there is a trend toward the double name "electrical engineering and computer science" to reflect the expanded discipline of these departments. Appendix B presents the curriculum of the undergraduate program in electrical engineering at the University of Tokyo, and Appendix C lists the equipment available at the computer centers and semiconductor laboratories process at the University of Tokyo.

Industry-university cooperation. The nature of industry connections to universities in the United States and Japan differs significantly. In the major engineering departments in the United States, faculty members are closely tied to companies through consultant relationships, memberships on boards of directors, and, in some instances, direct participation in the development of new companies. In Japan, faculty members and staff at national universities are paid twelve-month salaries and are regarded as government employees, which makes it impossible for them to accept fees or salaries as industry consultants. Yet despite these restrictions, there is a good deal of interaction between university personnel and industrial researchers in Japan, which often results in the channeling of students to particular companies.

In the United States, research grants available to graduate students have come primarily (more than 95 percent at Stanford) from the federal government, but the private sector provides some funds by way of corporate affiliate connections. These ties, which normally involve a modest annual contribution of unrestricted funds to a university (typically $5,000 to $20,000 a year per company), serve to facilitate communication but do not represent any significant corporate involvement in the day-to-day research (as contrasted to some of the new forms of industrial support for university R&D described on pp. 12-13). These programs ordinarily involve annual presentations of the results of research by graduate students, visits by industry personnel to universities, interviews with graduate students, and informal discussions of technical issues. Probably the greatest expansion of programs has occurred at Stanford, M.I.T., and certain other universities that are close to thriving industrial parks.

The growing importance of graduate training has led to the development of innovative programs to provide further education to personnel already employed by companies. Stanford's experience illustrates the evolution of such programs. In 1954 nearby companies began sending selected employees to the university, some of them on a part-time basis in the master's degree program, others as full-time students. The companies paid the full cost through tuition-matching grants. Such industrially supported graduate education programs still exist in many places across the United States. Some companies, like Bell Laboratories, routinely place every new employee with a bachelor's degree into a master's program during the first year of employment.

In the mid-1960's, as it became more difficult for employees to commute from their jobs to the campus, instruction by means of closed-circuit television was introduced. (At Stanford, the cost of establishing this system was borne by the cooperating companies.) In the early 1970's, experiments were done in videotaping the classes that were being transmitted to the cooperating companies via closed-circuit television. In this way, companies remote from the university could participate. This has proved successful in many instances; the tapes are played before small groups in the universities, with guidance provided by a qualified tutor.

Like their American counterparts, Japanese departments of electrical engineering have looked primarily to the government for financial support of their research programs. In 1981 the Ministry of Education allocated approximately $160 million in funds for materials and equipment. Graduate students depend either on family funds or on scholarships provided directly by government or private agencies; the Ministry of Education grants cited above are expended solely for materials and equipment. Industry provides a modest amount of direct support for research also.

Given their different approaches to innovation and to the preparation of new professionals, how do the U.S. and Japanese semiconductor industries compare with respect to state-of-the-art technologies? The next chapter considers the available evidence concerning the technological strengths and weaknesses of each country's semiconductor industry at the present time and the conditions that may affect the persistence of those strengths and weaknesses.

THREE

Technological Resources

*Franklin B. Weinstein, Michiyuki Uenohara,
and John G. Linvill*

Perhaps the most striking characteristic of the semiconductor industry
is its technological dynamism. The pace of technological advance in
this industry has been so dramatic that one must be exceedingly cautious when attempting to identify long-term trends in technological capabilities. The frequent introduction of new products and the shrink-
ing time lag between product generations make it unusually difficult to
project the future impact of current advantage. A discussion of present
technological strengths and weaknesses can only suggest what is hap-
pening and what seems possible if not probable tomorrow. One cannot assume the persistence of present patterns and trends, nor can one
take it for granted that present advantages can be maintained or that
current weaknesses cannot be overcome. In short, it is simply impossible to predict the winners in the competition for technological leadership in the semiconductor industry.

Any comparison of technological strengths and weaknesses of the
U.S. and Japanese semiconductor industries must be prefaced by a further caution. The Japanese semiconductor industry is relatively ho-
mogeneous, in that semiconductor production comes overwhelmingly
from large, vertically integrated companies, whereas the U.S. industry
includes vertically integrated companies, captive producers, and independent merchant producers. The heterogeneity of the U.S. semiconductor industry makes it very difficult to issue valid generalizations
about the two industries as a whole. What may be true for a small
merchant component producer may not apply at all to IBM. The
growth of cooperative relationships between U.S. and Japanese companies may further blur the distinction between industries in the two
countries.

In addition, we should keep in mind that it is extraordinarily difficult to obtain hard data from industry sources to support or refute press statements about the relative strengths and weaknesses of U.S. and Japanese semiconductor technologies. This chapter relies heavily on interviews and correspondence with industry executives in both the United States and Japan.

As the semiconductor industry moves into the VLSI era, it faces formidable technological challenges. The U.S. industry has in its favor substantial strength in basic research, new product design, and product application, and also a great degree of flexibility, reflecting the special capabilities of the huge systems house, IBM, as well as the many smaller but very successful entrepreneurially oriented companies. Japan has in its favor a high level of process technology, efficient mass production, and the ability to maintain quality and reliability reflecting the financial capabilities of the large diversified Japanese companies and the strengths of the Japanese work force. But in making these broad generalizations, one must recognize that the U.S. industry—certain companies in particular—possesses capabilities for high-quality production that rival or surpass those of the Japanese; similarly, in certain areas of new product development, the Japanese appear at this moment to hold an edge.

Following the opening caution, we need to be wary lest we enshrine as permanent what may be only a momentary advantage. The capacity of U.S. companies to respond to the Japanese challenge in quality mass production is formidable, as demonstrated by the substantial progress made in the last year. By the same token, it would be a mistake to conclude that Japan's educational system or industrial structure imposes any insurmountable obstacles to innovation. The overall technological supremacy of the United States in basic research and semiconductor product design is not at all beyond challenge by the Japanese.

Public discussion of U.S.-Japanese competition in semiconductor technology has tended to focus too heavily on the standard memory chips—the 64K and 256K dynamic RAMs (random access memories containing, respectively, approximately 64,000 and 256,000 memory cells)—which have been presumed to be important in amortizing the future development of the industry and are indeed the area in which the Japanese challenge has so far been concentrated. But it should be remembered that the competition between the U.S. and Japanese semiconductor industries is being waged not only in high density standard memory chips but also in logic chips, such as the microprocessor or computer-on-a-chip, customized chips, electronic systems utilizing semiconductor products as components, and many other products. Standard memory chips are very important, but they tell only part of the story.

The Development of New Semiconductor Products

There is no doubt that U.S. companies have played the preeminent role in basic research and the development of new semiconductor products since the invention of the transistor. Practically every major breakthrough in the semiconductor field has come from the U.S. industry. There is every indication that the U.S. industry will continue to pour substantial resources into research and new product design. Tax credits for R&D, and the revival of government funding for research, as evidenced by the VHSIC program and the DARPA supercomputer project, will provide a boost to the U.S. industry's research efforts. So will the establishment of university research centers jointly funded by industry and government, as discussed in Chapter 2.

These areas of basic research and new product development have generally been regarded as weak points in the Japanese semiconductor industry. The invention of the tunnel diode at Sony in the middle 1950's is said to be Japan's principal contribution to basic research in the semiconductor field. Some would say that this "weakness" has not been very costly to Japan, because the Japanese industry has benefitted from a "free ride" on U.S. R&D: American companies have borne the burden of designing new products and supporting the basic research that makes those products possible, while the Japanese, emphasizing the development of process technology that enables them to replicate or adapt U.S. designs at low cost and high quality, have often been able to capture a large share of the market in a relatively short time. The speed with which product generations occur in the semiconductor industry has made it very profitable to be first with new product designs. But as the time required to replicate semiconductor product designs has shrunk, the commercial advantages of being the design leader have diminished.

In any event, the Japanese are sharply aware of their deficiencies in basic research and product design, and they are determined to overcome them. In the first place, even if the commercial advantages of being first with a new product are less than they once were, those advantages are still significant. Moreover, Japanese engineers assert that technology transfer between Japan and the United States has been too one-sided. There is a growing reluctance on the part of some U.S. manufacturers to share their technology with the Japanese through licensing agreements, and the Japanese increasingly recognize that continued access to the fruits of U.S. research and development will depend, at least in part, on their offering enough to the U.S. side to make a fair exchange.[1]

Japan's government-supported project on VLSI research and the more recent program on optoelectronics (described in Chapter 2) represent a major effort on the part of the Japanese to boost basic research capabilities and to foster the development of new products and processes. Representative results of the VLSI project include:

1. A method of evaluating circuit design by computer
2. Test-pattern generation of combination circuit blocks by statistical method
3. Electron beam lithography machine using field emission cathode
4. Variable-beam type electron lithography machine
5. Electrical-mechanical, hybrid-type electron beam lithography machine
6. Software system for electron beam lithography
7. Mask pattern tester
8. Eight-inch-diameter silicon wafer technology
9. Defect-free, uniformly thin epitaxial layer growth technology
10. Optical pattern projection machines
11. Submicron pattern projection machines
12. Dry etching machines
13. Dry etching technologies for submicron patterns
14. VLSI material evaluation technology
15. Evaluation technology for oxide and nitride layers by liquid crystal
16. Device evaluation and measurement system
17. VLSI device computer-aided design (CAD) and simulation technology
18. New VLSI memory element design
19. Eighteen-bit Register Arithmetic Logic Unit of 400 ps
20. High-speed bipolar memory—4-kilobit CML random access memory (RAM)
21. 256-kilobit MOS RAM
22. 512-kilobit read only memory (ROM) using direct electron beam lithography

How significant are these results, in light of the work that has already been done in the United States? Asked to evaluate this list of Japanese accomplishments, a leading American expert on VLSI technology expressed the opinion that, except for the work using liquid crystal (no. 15), the Japanese did not appear to have made any major breakthroughs. In most areas, he felt that the Japanese had simply extended their technology in ways comparable to developments that had already occurred in the United States.

The 64K and 256K RAM competition. The determination of the Japanese to compete with the United States in the development of new product prototypes has been most apparent in the development of state-of-the-art VLSI memory chip technology—specifically, the 64K and 256K dynamic RAMs. The competition for market share in the 64K RAM, successor to the 16K RAM, is a test for producers of memory chips. Sales of the 64K RAM are expected to reach a peak of $1.8 billion by 1985. By 1989 the annual market for the 256K chip is expected to hit $3.7 billion. The Japanese have spent very heavily to develop and produce these chips, and this is the area in which the most dramatic Japanese successes have been achieved.[2]

Although IBM reportedly produced the 64K RAM in high volume well in advance of the competition, the first 64K RAM to reach the merchant market was Japanese, introduced by Fujitsu in 1978. This chip proved a failure in the marketplace because most customers preferred a single power source, rather than the dual source designed by Fujitsu. In 1980, Motorola's 64K RAM became the largest seller, but NEC, Hitachi, Fujitsu (with a new model), and Toshiba all began producing technically impressive 64K RAMs in large volume.[3]

By the end of 1981, the Japanese had captured 70 percent of the 64K RAM market, and some industry commentators declared that Japan had "won" the battle for the memory market. As one writer put it, "The struggle for dominance of a new generation of memory chips is already over, almost before the chips hit the marketplace."[4] The prospect of Japanese domination of the 64K RAM market was so alarming that senior U.S. government officials were said in February 1982 to be considering restrictions on the import of 64K RAMs in the interest of national security. American semiconductor producers were also reported to be exploring the possibility of filing antidumping petitions, charging the Japanese with predatory pricing of the 64K RAM, an allegation that the Japanese denied.[5]

In the spring of 1982, Japanese exports of 64K RAMs to the United States reportedly began to decline. Official Japanese sources confirmed a slackening of deliveries, but claimed that this was not a response to trade pressures; rather, delivery dates were pushed back because of high demand in the Japanese domestic market.[6] Led by Texas Instruments, which by mid-1982 had overtaken Motorola as the top U.S. merchant supplier of 64K RAMs, the U.S. industry appeared to be narrowing Japan's lead.[7] By mid-1983, the Japanese share of the 64K RAM market was estimated by the Semiconductor Industry Association to be about 55 percent (although a 70 percent figure was still widely cited in press accounts).

Estimates of market share reported in the press exclude captive production by companies like IBM and Western Electric, which produce RAMs only for their own use. (Western Electric, however, announced plans to sell chips on the merchant market beginning in late 1983.) As of the end of 1981, according to an IBM company source, IBM had "shipped more 64K RAMs than the rest of the world combined, probably by an order of magnitude," and IBM remains the largest producer of 64K RAMs in the world. Indeed, in mid-1983 captive chipmaking—led by IBM with a projected IC production value of more than $1.2 billion in 1983—had become the fastest growing segment of the industry.[8]

One reason why the Japanese were able to enter the 64K RAM market so early and with such strength was their adoption of a conservative design. The Japanese used as a base for their 64K RAM the cell design originally used in Mostek's 16K RAM, improving on it and, in effect, scaling up the techniques that had proved successful at the 16K level.[9] Japanese engineers claim that the more conservative design of their chips makes them more reliable, less apt to be defective. Memory layouts in the Japanese 64K RAM tend to be significantly larger than those of U.S. chips—about 50,000 square mils (a mil is one-thousandth of an inch) as compared with 34,000-45,000 square mils; the smaller U.S. design is intended to reduce production costs by getting more dice on a wafer.[10] Designing larger cells in larger dice helps the Japanese avoid "soft errors" (erasures caused by radiation emitted by the chip's ceramic package) in their 64K RAMs.[11]

Except for Motorola, which, like the Japanese companies, chose a conservative design and established a strong early market position, U.S. manufacturers have tended to pursue more complex designs. Texas Instruments' chip has a more complex design, works more quickly, and sells at prices up to 30 percent higher than slower chips. In March 1982, one industry analyst pronounced Texas Instruments' 64K RAM "the best chip in the world."[12] Shortly thereafter, Intel, which had previously announced and then withdrawn a 64K RAM, introduced a new 64K RAM with the smallest design size (i.e., higher yield and lower cost per part) and the highest capacitance (i.e., increased immunity to errors in storing information).

Some analysts think that Intel's 64K RAM leapfrogged the competition, not only because of its small size and high capacitance but also because of its use of redundancy, which provides an extra margin for error. Intel has applied the principle of redundancy to the 64K RAM by incorporating 3,000 spare cells (creating, in effect, a 67K chip without increasing memory capacity) and dramatically increasing yield by using those spares to replace cells that prove defective. Redundancy has been used by both Intel and IBM in their magnetic bubble memories. The Japanese are undecided about its merits. One Japanese executive described redundancy as a stopgap measure rather than a genuine solution to the 64K RAM's problems, but NTT and NEC are said to have used redundancy in their prototype versions of the 256K RAM. Whether the principle will be adopted as an industry-standard design is uncertain.

Intel suffered from various delays, but by mid-1983 its redesigned 64K RAM had begun selling well, and Intel's share of the 64K RAM market was expected to rise from a mere one percent in 1982 to 3-5 percent in 1983 (still far below Intel's 15-20 percent share of the 16K RAM market). Another major U.S. chip producer, National Semiconductor, ran into problems when its 64K RAM, which sought to use a complicated triple-polysilicon process to reduce the size of the individual chip, found it could not manufacture the chip economically in high-volume production. National switched plans and opted for a design from a Japanese company, Oki, and volume production was expected by late 1983.[13]

What are the implications of Japan's early lead in 64K RAM sales? Being first to market with a new product has always been important in the semiconductor industry, because those with the highest volume sales early are in a position to retain their market share by cutting prices more rapidly than their competitors can afford to do. Japanese 64K RAMs sold for $28 in March 1980; by the end of 1981, the price had dropped to $8, and by April 1982 to $5-$6. These low prices deprived latecomers of the early profits needed to finance new products.[14]

On the other hand, the enormous market for the 64K RAM makes it possible to overcome much of the disadvantage resulting from a late entry. Indeed, surging demand for the 64K RAM in mid-1983 pushed prices up, opening new opportunities for latecomers to gain a reasonable return on the chip. The Japanese captured 70 percent of a 64K RAM market that totaled $100 million in 1981, but that was very early along the road to the $1.8 billion market predicted by 1985. As noted, their share had dropped to about 55 percent by mid-1983. As one analyst put it, "the memory market is not a 100-yard dash; it is a marathon, and the early foot is not necessarily among the winners." Texas Instruments, following several failures, increased its market share with its highly acclaimed chip. The same analyst noted that a company like Intel, with a superior product, could "enter late and win plenty of business because there is not enough capacity to satisfy the demand." Besides, Intel's superior design reduced unit cost, enabling Intel to keep its prices competitive. As of May 1983, Mostek, the leader in 16K RAM sales, was shipping more than 3 million 64K RAMs monthly, and was said to be well on the way to becoming the leading U.S. supplier of 64K RAMs.[15] But late entry means that some customers are already tied into agreements with other suppliers, and companies like Intel and National will not be able to capture the market share they could have had with an earlier entry.[16]

Japan's share of the 64K RAM market is expected to decline further. But since they held 40 percent of the market for the 16K RAM, the Japanese are considered unlikely to end up with much less than 50 percent of the 64K RAM market.[17] This would of course be a significant achievement for the Japanese chip makers. Still, the 64K RAM experience should serve as a reminder that the early returns do not tell the whole story. Although the Japanese may in some sense have "won" the battle of the 64K RAM, it is not the total disaster for U.S. manufacturers that early press accounts indicated.

Some analysts expected the Japanese to be in a particularly strong position with regard to the 64K RAM's successor, the 256K dynamic RAM. By 1982, prototype versions of the 256K chip had already been built and tested at NTT's Musashino Laboratory (the cooperative facility established as part of the government-sponsored VLSI project), NEC, Hitachi, and Mitsubishi. By late 1981 Oki Electric was reportedly sending samples of its 256K RAM to potential buyers for testing, and Fujitsu was believed to be close to doing so. Japanese press reports indicated that Hitachi would be the first to mass produce 256K RAMs. A Hitachi spokesman denied that there were any concrete plans for early mass production, and an NEC executive asserted that Hitachi would be "committing suicide" if it started selling 256K RAMs in 1983, since this would "kill the 64K RAM market before it got off the ground." NEC itself is expected to start producing 256K RAMs by 1985 at its planned facility in Roseville, California.[18] By mid-1983 six Japanese companies (the above four, plus Toshiba and Mitsubishi) were shipping sample 256K RAMs to their U.S. customers for evaluation, and commercial production was anticipated for 1984.[19]

The Japanese will undoubtedly be very strong in the 256K RAM market, but again, the central point is that a strong early showing does not preordain the long term outcome. In fact, many Japanese (and Americans) were surprised when Western Electric announced in May 1983 that it would ship 256K RAMs to its customers in large quantities during the last quarter of 1983, ahead of the Japanese.[20] In 1982 IBM had announced a 288K dynamic RAM, which according to a company source had already been produced on a production line, and IBM was expected to announce a new line of mainframe computers incorporating 256K chips by the end of 1983.[21]

The surprise emergence of Western Electric as the early leader in the 256K RAM competition illustrates the volatility of this industry. Notwithstanding the strong Japanese showing on the 64K RAM, there is no way to predict who will ultimately dominate the 256K RAM market, expected to reach $3.7 billion by the end of the decade.[22] There are several important differences in the circumstances surrounding the 64K and the 256K RAM competition.

First, the 64K RAM emerged amid a severe recession, which imposed a particularly severe handicap on U.S. companies without the enormous financial resources of the Japanese manufacturers. The recession also limited demand, which facilitated price cutting by the Japanese. With demand rising rapidly, prices are likely to hold, and with the economy in an upswing, the U.S. companies are likely to be in a stronger position. Second, the 64K RAM chips, particularly those of Japanese design, relied on relatively well known manufacturing techniques. The 256K RAM will require more innovative approaches. Certain U.S. companies, like Intel and National Semiconductor, suffered in the 64K RAM competition because the new technologies they had developed proved difficult to translate into cost-effective mass production. But the experience they gained on the 64K RAM may give them an edge with regard to the 256K RAM. Thus, one leading industry analyst in mid-1983 predicted "a very balanced battle for the next two years," and another, who anticipated a Japanese victory, noted that even the "losers" would thrive because "the whole pie" was growing so fast.[23]

In the field of static RAMs, which are technologically more complex than dynamic RAMs because they do not need to be electronically refreshed, Japan has apparently taken a lead. Matsushita was the first company in the world to describe a 64K static memory, and Hitachi and Toshiba, as well as Matsushita, were gearing up to produce 64K static RAMs, at a time when, according to one observer, U.S. chip manufacturers were "barely getting their 16K static RAMs out the door."[24]

Areas emphasized by the Japanese. Japan's most impressive technological achievements in basic research and new product development have been in the areas of optoelectronics, gallium arsenide circuitry, silicon-on-sapphire (SOS) technology, displays (both liquid crystal and light emitting), and high frequency transistors.[25] According to some analysts, the Japanese industry presently has an edge in all those areas. Specific examples of new products designed first by the Japanese include: high electron mobility transistors (HEMTs), developed by Fujitsu; long wave length semiconductor lasers (1.3 and 1.5 micron); and optical fibers with very low loss (0.2 decibels per kilometer). In addition, many special purpose designs have been developed for in-house use, such as completely new concepts of LSI for pattern recognition, but these have not been disclosed publicly. There are ongoing basic research projects concerning digital gallium arsenide circuits and Josephson junctions, both aimed at producing very high speed integrated circuitry, at NTT's Musashino Electrical Communication Laboratory (Japan's counterpart to the Bell Laboratories), and at Fujitsu, Hitachi, NEC, and Mitsubishi. The Tokyo Institute of Physical and Chemical Research and Tokyo University both have major programs devoted to the development of Josephson junction technology.

Gallium arsenide technology is being developed in the United States as well, but these efforts are aimed mainly at military applications. Furthermore, gallium arsenide development is being undertaken primarily by large systems houses for their own internal use, not by companies that are likely to become merchant suppliers of gallium arsenide components. The Japanese, on the other hand, are developing gallium arsenide technology specifically for application to potentially high-volume consumer products, such as satellite-to-home television receivers and intrusion alarms for home security using radar techniques. Japanese companies can be expected to become merchant suppliers of gallium arsenide components.

In both the United States and Japan, some makers of mainframe computers have come to view gallium arsenide integrated circuit technology as a desirable means of achieving higher speed and density. Companies such as IBM, Burroughs, and Sperry-Univac have been developing this technology, but the most advanced work in the application of gallium arsenide ICs to mainframe computers is being done at Fujitsu. If the trends continue, Fujitsu is likely to emerge as the world leader in large mainframe computers using gallium arsenide IC technology.

For several years Fujitsu has been supplying central processing unit (CPU) gate arrays to Amdahl for all their mainframe computers. Some observers think that Fujitsu may develop an all-gallium arsenide main-frame computer with speed and processing capabilities far surpassing those attainable with silicon. The Josephson junction technology developed by IBM has comparable capabilities, but many problems need to be resolved before a complete mainframe computer using Josephson junction technology can be constructed. In June 1981 some observers in Japan reported that the Japanese government was apparently preparing to launch a drive to develop the most advanced Josephson junction circuits yet. Government programs described in Chapter 2 on optoelectronics and software development seem likely to ensure that the Japanese technological challenge will become increasingly formidable.

It is important to note, however, that some of the areas in which the Japanese appear to have taken a technological lead are areas that the U.S. industry has chosen not to enter because it emphasizes other areas or is working on different approaches. Gallium arsenide technology, for example, has been a matter of some controversy, and many American specialists believe that its attractiveness as a means of achieving high speed is more than offset by insulation problems that make gallium arsenide a much more difficult material to deal with than silicon oxide. Gallium arsenide also poses difficulties with respect to doping and diffusion. No major computing system presently uses gallium arsenide, and it is not at all certain that the problems connected with the use of gallium arsenide can be resolved. An IBM executive expressed doubt that gallium arsenide would be extensively used in high performance machines very soon, and he predicted that even if Fujitsu did succeed in putting one into being, it would be "a curiosity piece." He saw far greater possibilities for gallium arsenide as a technology for communication and display application, and he noted that the Japanese had focused considerable attention on this particular application.

The broad base of U.S. supremacy. If the Japanese are challenging U.S. technological leadership in memory chips, optoelectronics, gallium arsenide technology, and the several other areas outlined above, U.S. companies seem to be still well ahead in virtually all other aspects of semiconductor product technology—especially in complex logic systems, microprocessors, and software capabilities. Both U.S. and Japanese engineers acknowledge that the U.S. industry presently holds the advantage in the design and development of systems that use microprocessors, in most aspects of microprocessor and computer technology, and in the customization of chips and other specialized user applications of semiconductor technology.

The U.S. lead in these areas may be attributed to three interrelated factors: (1) the U.S. industry's traditional strength in basic research and frontier technology, already discussed; (2) the superiority of U.S. engineers in software capabilities and in the automation of design; and (3) the structure of the U.S. industry and the nature of the U.S. market, which have encouraged microprocessor development and the design of products suited to specialized user needs.

The growing importance of software in the design of new semiconductor products makes the U.S. industry's advantage in this area particularly significant. Until recently, though the potential growth in number of elements per memory chip is essentially a matter of hardware, one of the main barriers to further development of more complex logic systems was software design. Hardware costs have declined at a spectacular rate, but software costs have not; indeed, in system after system, software costs have proved unexpectedly large. In the more complex systems, design costs are actually increasing per unit of element count. One report estimated that about 80 percent of the cost of developing new systems went into software, which led that writer to conclude that software support would be "the making or breaking point of market success in the 1980s."[26]

The U.S. lead in software is also an advantage in the development of a highly competitive position in the design of customized and semi-custom chips. Whereas some in the industry question the long-term potential for custom chips, others foresee a booming market for these products. A growing interest in customized chips is said to have been facilitated by improved technology, which makes it possible to squeeze tens of thousands of logical elements—known as gates—onto a single chip. The key to success in customized chips is quick design turnaround time, and technological progress in the automation of design has made customization more practical. The custom chip makes possible a variety of specialized applications that give a company's product unique features and enable a small company to carve out a niche in the market. Furthermore, though standard parts will continue to be important, ever rising densities inevitably push toward greater emphasis on customization. As density approaches a million electronic elements per chip, it makes sense, according to one industry expert, "to design almost all complex chips as microprocessors or as gate arrays or other semi-custom parts, and to produce perhaps five percent of them as full custom designs." Since a million-element chip is in sight, it is said, "customized circuits seem sure to become the standard parts of the future."[27]

The market in the United States indicates a much greater demand for microprocessors and for specialized applications than in Japan. One Japanese engineer described the U.S. market as more forward looking in this respect. In any case, the greater American market for these products is said to have been an important incentive stimulating their more extensive development in the United States. Furthermore, the existence of smaller entrepreneurial units in the United States facilitates the development of specialized user applications, because these smaller companies, though hardly competitive against the giants in the mass memory market, have the flexibility and motivation to tackle narrowly defined markets. So far as U.S.-Japanese competition in this line is concerned, U.S. companies have a distinct advantage in the U.S. market at least, because customized chips require close and continuous relations and the language barrier alone is formidable to the Japanese. Furthermore, spin-off companies that are formed to deal with some specialized need that a larger company disdains because of its narrowness are extremely rare in Japan. Big companies are often geared mainly to serve big customers, and small makers of customized chips can do well serving customers that larger companies do not wish to serve.

Japanese efforts in software and logic systems. Most of the Japanese engineers interviewed predicted that the U.S. lead in logic systems would continue for some time. They asserted that the basic architecture and ideas probably would continue to come predominantly from the United States, provided the U.S. industry has the funds for R&D. There was some suggestion, however, that the Japanese were adopting U.S. designs mainly to facilitate penetration of the large U.S. market, not because those designs were inherently superior. A Toshiba engineer expressed confidence that Japan would not equal the United States in software capabilities in the next five years, because "it is very difficult to organize software." The problem of developing Japanese-language software represents an additional impediment. But it would be a mistake to conclude that the Japanese have decided to accept a subordinate position in logic systems, or even in software. On the contrary, they are making a concerted effort to overcome their weaknesses in those areas. Mention has already been made of the MITI-sponsored program for software development. A major project on LSI design automation at the Electrical Communication Laboratories of NTT will also make a significant contribution to this effort.

Indeed, a 1982 report, based on a visit to Japan by Jerry Werner, the editor of *VLSI Design*, a technical journal, raised some doubt about the common assertion that the Japanese were far behind in software and in the automated design necessary for success in building custom ICs. Werner's report suggested two main reasons for the prevalent view that the Japanese lagged badly in software. First, with the exception of Fujitsu, Japanese semiconductor makers generally did not actively market custom and semicustom ICs in the United States; even in Japan, the software and hardware used to develop these devices often was unavailable outside the company. This meant that Japanese technical capabilities in this area were not highly visible in the United States, but it did not mean that they did not exist. In fact, there was substantial development of customized chips in Japan (indeed, custom chips predominated over general purpose chips) but these were mainly for in-house use. Second, the Japanese, in their hunger for information, regularly bought "one of everything" in design software. Some Americans mistakenly took this as evidence of Japanese weakness.

Werner found "brisk" activity in Japan in all areas of computer-aided design (CAD) and design automation (DA): high-level languages, automatic layout, circuit and logic simulation, and logic and layout verification. He was especially impressed by Japanese progress in layout verification and noted that the Japanese companies he visited displayed a uniformly high level of competence, even though they did not attain the degree of expertise found in Bell Laboratories or IBM. He observed that the Japanese emphasis was on evolutionary, not revolutionary, development of CAD and DA tools and systems.[28]

Software has had the reputation of being an area of low status in Japan, but the field's prestige is said to be growing, and the number of Japanese software engineers is increasing rapidly. The diffusion of microcomputers to high school and junior high school children and the popularity of "bit" shops are said to be indicators of the growing interest in software. Though some of the Japanese engineers interviewed blamed Japan's deficiencies in software on the failure of Japan's very "standardized" educational system—and even of Japanese culture generally—to encourage creativity and independent thinking, Japanese educators insist that this has changed. Japanese schools not only develop very high level skills in math, science, and technical matters, but they make a conscious effort to foster creativity.

With respect to formal training in software, there have also been important changes in Japan. The University of Tokyo has established an interdepartmental course of study in computer science, which functions as a department in the graduate school, and other major national universities have established departments of computer science. A great deal of software training is carried out by Japanese companies also.

The theme that Japan's weakness in software has been exaggerated continues to be sounded by analysts.[29] Indeed, it may well be that the principal difference is not in the quality of U.S. and Japanese software but in the approaches taken by the two countries. The organizational structure of the Japanese software industry has tended to obscure the significant work being done by the Japanese, leading to the erroneous conclusion that the Japanese are incapable of impressive software development.[30]

In the United States, independent software houses developed widely acclaimed packages. The Japanese emphasized customized software development, undertaken mainly by users and secondarily by computer manufacturers. Thus Japanese software was made less visible. Japanese software in certain fields (e.g., banking and robotics) has been said to be unsurpassed by anything designed in the United States.[31]

There was, of course, an obvious weakness in the Japanese approach—namely, that it required an enormous number of software developers. Recognizing this weakness, the Japanese government began in the early 1980's to channel funds to independent software houses (which, unlike their U.S. counterparts, cannot look to venture capitalists for support).[32] Clearly, it would be a mistake to take the perceived Japanese weakness in software development as a reflection of any intrinsic differences on the part of the Japanese. Just as the Americans were probably stronger in memory chips than was commonly believed, Japanese weaknesses in software development appear to have been exaggerated.

Besides improving their software capabilities, the Japanese are by no means out of contention in certain aspects of logic system and microprocessor development. As indicated, the Japanese are in a strong position with respect to high-speed integrated circuits. Many analysts believe the Japanese are ahead in high-speed bipolar gate arrays; this is said to have been a principal reason for the joint venture between Fujitsu and Amdahl, with the Japanese firm supplying gate arrays for its American partner's high-speed computers (although it should be recalled that Fujitsu entered the bipolar gate-array field as a second source for Motorola, not as the developer of its own technology).

Although the U.S. companies still hold a huge lead in microprocessor sales, Japanese companies have developed some advanced microprocessors. NTT announced in late October 1981 that it had developed Japan's first 32-bit microprocessor, only months after Intel, Bell Laboratories, and Hewlett-Packard had announced their own designs. Several Japanese companies are said to have devised special-purpose microprocessors for in-house use. An Intel executive indicated that he regarded NEC as his company's most formidable competitor in the microprocessor field. Still, the Japanese microprocessor leaders—NEC, Hitachi, and Matsushita—built respectable market shares only at the relatively low-value, low-power end of the microprocessor market. Japanese companies selling the more advanced microprocessors have for the most part copied or licensed U.S. designs.[33]

Japan's success in memory chips must be viewed in relation to MITI's longer-term goals for the computer industry. It is true that the skills that make for excellence in memory chips are not necessarily transferable to logic systems. The design of complex logic systems is very time consuming and expensive; it is not as straightforward as the design of memory chips, which are highly regular and conceptually simple. But companies like Fujitsu, Hitachi, and NEC are computer companies, not just makers of semiconductors or communications equipment. Though most of their exports to the United States have been memory chips, they are working hard on gate arrays, customized chips, packaging, and analogue and special circuits. Memory chips may be easily designed but they are very demanding in lithography and feature size; much of the learning and technology is transferable to custom ICs, logic, microprocessors, and so on. In short, we must assume that for the Japanese, memory chips have been merely a means to an end.

Whether the overall advantage that the U.S. industry has held in the development and design of new semiconductor products can be sustained in the face of a growing Japanese technological challenge is one question. Most of the engineers interviewed felt that with adequate financial and human resources, the United States probably would be able to maintain an edge, but it would be unrealistic to assume that the U.S. industry can recapture the degree of technological predominance it has traditionally enjoyed. From a commercial standpoint, however, that is not the only question. The development of process and manufacturing technology is just as important as the capacity to design new products. It is in this area that competition between the United States and Japan promises to be most intense.

Process Innovation, Production Technology, and the "Quality Issue"

Japan's most impressive technological accomplishments have undoubtedly been in process innovation and the improvement of productivity and quality levels for mass-produced semiconductor products. Much of the research undertaken in the government-sponsored VLSI project concerned the improvement of manufacturing processes—for example, electron beam lithography, probably the project's single most important area of accomplishment. But it is important to remember that practically all the basic process innovation breakthroughs in the semiconductor field have come from the United States—including electron beam lithography, ion implantation, and plasma etching.

There are, to be sure, several completely new processes that were developed first by the Japanese, including uniform film epitaxial processes, highly pressurized oxidation (pioneered by Mitsubishi), low-temperature passivation (developed at Hitachi), and NEC's anodization process. Perhaps the most impressive invention is a machine developed by Toshiba that uses a highly focused beam of ions to trace circuits on silicon chips. This ion-beam machine, capable of drawing lines so fine that Toshiba expected to be able to increase the number of electronic devices on a single quarter-inch-square chip from 64,000 to about 4 million, was said by one leading U.S. analyst to give the Japanese a three-year lead over American companies in the manufacture of highly dense chips.[34] For the most part, U.S. companies have not used these processes because they have chosen to follow other approaches. Japanese engineers who were interviewed felt that the U.S. companies had something of a psychological reluctance to use foreign technology, and also disliked having to pay royalties, but they did not claim that their process technology innovations, such as those mentioned above, were superior to the alternative technologies developed by American companies. They only said that these processes were evidence of Japan's innovative capabilities.

The quality issue. Some of the Japanese engineers interviewed did assert, however, that the design of Japanese products was more conducive to the maintenance of high standards of quality. One executive claimed that Japanese companies undertook more evaluation and collection of information before sending a product to the mass production line; though, in his opinion, the larger U.S. companies also did this, smaller ones could not afford to do so.

It was frequently asserted, as noted in our discussion of the 64K RAM, that the Japanese generally were more conservative in product design. According to J. M. Juran, the Japanese tend to de-rate components—that is, they use components at less than their specified operating range, thereby reducing the chances of failure. For example, an electric light bulb specified to operate at 110 volts would have an average failure rate of x hours, but if you use it at a lower voltage—that is, de-rate it—it would have a much longer life.[35] In the 16K RAM, for example, the early American products were plagued by poor product yield and by "soft errors" (one-time errors that do not damage the structure and are therefore difficult to detect). The more conservative design adopted by the Japanese resulted in acceptable memory products from the first commercial shipment and rapid improvement of product yield as the quantity of production increased.

Japanese engineers described three design features to illustrate how they had "designed quality into the chip." First, they were more liberal than their U.S. counterparts in the use of border areas; this was expensive, but by leaving more space, they were able to avoid certain bonding problems. Second, they claimed to have made an important breakthrough in the packaging process. Heretofore, the two choices in packaging had been hermetic, which is very reliable but expensive, and plastic. The Japanese developed a new plasma nitride passivation technology that made plastic more reliable. (According to the Japanese, American experiments along the same line produced products that were too thin and often defective, not comparable in quality to the Japanese packaging.) Third, the conservative design of Japanese chips was said to have made it easier to deal with the problem of alpha particle immunity, which was a cause of soft errors. The Japanese developed a technology to overcome the alpha particle problem by overcoating the chip.

Just how significant these design innovations are depends on one's point of view. As an American executive observed, these are but three features out of several hundred that need to be considered in any assessment of chip design. Some U.S. executives were not even sure that the Japanese should be given credit for a pathbreaking role in nitride passivation technology; it was noted that an American engineer then at Texas Instruments earned most of the key patents in this area. To be sure, some differences between the U.S. and Japanese chips may have existed in the early days of the 16K RAM, but even in the three features just described, there was no longer any significant difference between the U.S. and Japanese designs.

The issue was one of timing. Most early U.S.-made 16K RAMs were plagued by alpha particle problems, and the Japanese avoided these because of a more conservative design and a better ceramic package. If the early Japanese chips were better designed from the standpoint of the alpha particle problem, this was because of requirements imposed by NTT. But Intel, Texas Instruments, IBM, and Bell all designed remedies for the alpha particle problem in the form of multifaceted design fixes. American and Japanese engineers came to similar discoveries in plasma nitride passivation technology at about the same time. Similarly, the Japanese may have been more liberal in the use of border areas, but that is no longer the case. On balance, it seems difficult to sustain the claim that Japanese 16K RAMs are superior in design.

Much of the Japanese reputation for superiority in process technology has been based on the belief that the Japanese companies, through a high degree of automation and the effective use of techniques for quality control, have produced very high quality semiconductor devices. There is some evidence that for a limited period of time prior to early 1981 Japanese-made 16K RAMs achieved substantially higher overall quality levels than their U.S.-counterparts.

The most widely publicized testimony to the superior quality of Japanese-made 16K RAMs was that of Richard W. Anderson, general manager of the Data Systems Division of Hewlett-Packard, speaking at a seminar in March 1980. In 1977 Hewlett-Packard began importing 16K RAMs from Japan because American suppliers, which had cut back investment during the 1974-75 recession, could not produce enough chips to meet the company's demands. According to Anderson, Hewlett-Packard, over a period of years, found that the Japanese memory chips were consistently of higher quality than the U.S.-made products. In tests of approximately 300,000 memory chips—half from three American suppliers, and the other half from three Japanese suppliers—there was a striking disparity between the Japanese and U.S. products. None of the Japanese lots was rejected because of failures, whereas failure rates for lots from the U.S. companies ranged from 0.11 to 0.19 percent. The simulated field failure rate of the best Japanese supplier was 0.01 percent failures per thousand hours, and the failure rate of the worst Japanese supplier was 0.019 percent per thousand hours. The best American supplier had a failure rate of 0.059 percent, or approximately six times as many failures as the best Japanese supplier, and the worst American supplier had about 27 times as many failures per thousand hours.[36] Although later studies by Hewlett-Packard showed significant improvement by the U.S. companies (see page 55 below), this early report had a dramatic effect in drawing attention to the issue of Japanese qualitative superiority.

Hewlett-Packard's findings were to some extent corroborated by data from other sources. An informal survey conducted by Benjamin Rosen, a leading semiconductor industry analyst, concluded that both users and manufacturers of ICs in the United States agreed that Japanese levels of quality were superior, though perhaps not so markedly superior as Hewlett-Packard's statistics indicated.[37] Roger Dunn, manager of the LSI/memory component engineering department at Xerox, said that his company's experience paralleled that of Hewlett-Packard; though the tests were very complicated, and it was difficult to be confident about the accuracy of minor variations, there was a pronounced pattern. Through late 1980, Japanese 16K RAM "average fallout" at Xerox's incoming inspection was typically in the 0.14 to 0.7 percent range, compared with 1.3 to 2 percent for U.S.-made devices. In a reliability test—a 96-hour dynamic burn-in at 125 degrees centigrade—Japanese RAMs showed failure rates of only 0.2 to 0.4 percent, compared with 0.3 to 1.2 percent for the U.S.-made 16K RAMs. In mid-1980 Xerox set maximum failure rate requirements that three of the company's four Japanese suppliers felt they could meet with their standard product, while the two principal American suppliers set up high reliability lines to meet those levels, and the other U.S. suppliers indicated that they could not meet those requirements on a monthly certified basis.[38]

The response in the U.S. semiconductor industry was defensive. Executives of several U.S. firms insisted that their users had found no difference in field failure rates between U.S. and Japanese memory chips. They questioned the validity of Hewlett-Packard's statistics and even suggested that the company had used improper screening procedures or had used a particular test that Japanese companies had adopted but U.S. companies had not.[39] Other U.S. industry spokesmen accepted the validity of Hewlett-Packard's survey, but considered it misleading. First, they contended that the Japanese had engaged in selective shipping, in which important U.S. customers like Hewlett-Packard received products that had been carefully screened. Second, they accused the Japanese of "quality dumping"—that is, maintaining artificially high prices in their protected domestic market that allowed them to subsidize their exports by spending more money to ensure a high degree of quality. In other words, even if Japanese chips were sold in the United States at a price comparable to that of U.S.-made chips, the Japanese products were said to be underpriced relative to the cost of producing such a high quality product.[40]

Whatever role selective shipping or quality dumping may have played in the establishment of Japan's reputation for high-quality 16K RAMs, there is something approaching a consensus that a major reason for Japan's success is to be found in the high priority that the Japanese have attached to quality and the effectiveness of Japanese programs for quality control. Many Japanese would also argue that the accelerated pace of automation in Japan is an important part of the explanation, but on this point there has been a good deal of controversy.

Certainly some U.S. companies have maintained quality levels equal or superior to those of the Japanese, but most observers would agree that many U.S. manufacturers have produced their semiconductor devices to meet contractual requirements that were competitively derived. In other words, the U.S. companies chose to build devices to "acceptable quality levels" not because they were incapable of higher levels but because, in the competitive market, there was often no need for, nor any reward for, higher quality. Of course, when Japanese companies or any other competitors achieved over a sustained period an even higher level of quality while preserving state-of-the-art technology and a competitive price, the competitive situation changed and all other competitors had to meet the challenge. As a Hewlett-Packard executive put it in early 1981, Japan had acquired a quality edge because U.S. management had not made a decision to take the lead in quality, but he thought that attitude was changing.[41]

Why did the Japanese decide at such an early date to produce memory chips of unusually high quality? The role of NTT as a major consumer of semiconductor memories was probably crucial. In the early 1970's, NTT's need for large quantities of 16K RAMs was a major stimulus to the Japanese semiconductor industry, and NTT established quality standards that were very high even for Japan, where high quality is expected. Quarterly reports to NTT's suppliers kept them informed on the performance of their products, and the suppliers did all they could to meet NTT's rigid specifications even when the costs were exorbitant, because they knew this would pay off in the long run.

It is often said that Japanese customers generally demand quality standards that many Americans would regard as unreasonably high. The particularly high standards enforced by NTT accentuated that trait and had a major impact on the development of Japanese memory chips. Most of the main Japanese semiconductor manufacturers have had long-standing commitments to NTT and to the communications industry in general, and other companies, not heavily involved as suppliers to NTT, have had to maintain comparably high quality standards in order to hold a small share. As the industry developed, the standards set by NTT came to apply to the industry as a whole. Companies that were unable to maintain those standards at low cost simply had to withdraw from the very large and competitive Japanese market.

In the United States, the high specifications set by the Department of Defense and by the space program played a similar role at an earlier

stage. AT&T, IBM, and other manufacturers of systems set high qualification requirements for their vendors, when they purchased components from outside the company, and this had an impact on the quality
of semiconductors. The U.S. telecommunications industry undoubtedly had quality requirements as high as those established by NTT.
There was a fundamental difference, however, in that Western Electric
produced its own high-quality memory chips for AT&T rather than
buying substantial quantities on the merchant market, as did NTT.
The impact of AT&T's high standards was not as far-reaching as that
of NTT's.

Another reason, often mentioned, for the superior quality of the
Japanese product is the overall approach to quality control, which is
fundamentally different from that practiced in the U.S. semiconductor
industry. The Japanese emphasize what they call Total Quality Control
(TQC) aimed at building in quality at every stage of corporate activity
and at every level of management and labor. In Japanese firms, quality
control is an integral part of the company's operations, involving every
employee. It is, as one Japanese semiconductor industry executive described it, *preventive* quality control, not detective control. Many U.S.
firms, though obviously not all of them, are said to look upon quality
control as mainly the responsibility of quality control experts—people
whose principal function is the weeding out of faulty products
through extensive inspection and testing. Where American quality
control procedures have focused on spotting and discarding or repair-
ing defective products, the Japanese say they have put more emphasis
on the need to understand failure mechanisms and remove them from
the production sequence.

To illustrate the Japanese commitment to preventive quality control,
the manager of a U.S. company that bought large quantities of semiconductor products compared his experience with U.S. and Japanese
companies when they were informed of a problem with the products
they had shipped to his company. The U.S. company sent a sales representative who spent a day at the company and then reported back to
his sales manager that his customer was angry; the sales manager invited the dissatisfied customer to lunch, where he offered assurances that
his company attached great importance to this business relationship
and indicated that he was prepared to take back the defective parts.
But, as the complaining executive put it, "nothing got fixed." The
Japanese company, upon being told of the complaint, dispatched a
team of technicians, armed with tools and clipboards, who "went
through everything with a fine-toothed comb." Shortly after their re-

turn to Japan, they sent a comprehensive report listing changes that the U.S. buyer needed to make, things that the Japanese supplier needed to change, and problems that might have come about in shipping. Through these preventive measures, the problem was solved.

One cannot draw definitive conclusions from individual accounts such as the above: attitudes and approaches are likely to vary from one company to the next, and what one U.S. executive described as a typical difference between U.S. and Japanese companies reflects only a limited range of experience. Certainly, however, the Japanese seem to place a very strong emphasis on quality. Japanese executives are said to go to extremes in exhorting their employees to strive for higher standards. In insisting on a commitment to zero-defect production, a Japanese manager told his employees at the morning rally that one faulty product in 1,000 might represent a failure rate of only 0.1 percent, but that for the customer who bought that product the failure rate was 100 percent; hence there must be no failures. The extensive use of quality control circles in Japanese companies is well known, as is the offering of incentives in the form of prizes for outstanding achievements in the quality control field. The overall effect is to create substantial positive peer pressure, which, in the words of one U.S. executive, has produced "tremendous work habits."

Whereas the commitment to quality control tends to be pervasive in Japanese firms, in many U.S. firms, according to U.S. industry sources, quality control efforts can often be hampered by conflict between those with production responsibilities and those in charge of quality assurance. This is particularly evident in the problem known as "end-of-the-month-itis." Companies that use semiconductor products as components need a constant linear flow of shipments, but many component manufacturers find themselves working overtime to ship the entire month's quota at the end of the month instead of spreading shipments evenly over four weeks. This problem probably occurs to some extent in Japanese plants as well, but it is said by U.S. and Japanese executives alike to be more severe in the case of the American companies. In part, this may reflect bottlenecks arising from the heavier dependence of U.S. companies on overseas assembly, as will be discussed later.

In the end-of-the-month scramble to meet production commitments, the priorities of production supervisors can clash with those of quality control personnel, whose insistence on meeting quality specifications can cause production quotas to go unfulfilled; if production quotas are not met, contracts may be lost. In such conflicts between

quality control and production personnel, the production people usually emerge the winners. Under pressure, the quality control or assurance personnel often ease up on standards, and products that
would have been rejected earlier in the month are allowed to pass in
order to meet production commitments. The position of quality assurance personnel within a company is not an enviable one. "Everyone
hates your guts in q.a.," one inspector said. "People get mad at you if
you reject their stuff." Quality assurance personnel regularly find
themselves accused of "holding up production."

On the other hand, most U.S. executives denied that they considered
quality control merely a process to weed out defective products. They
insisted that they tested at all stages, including the development stage,
and made every effort to prevent future failures. They even pointed out
that quality control techniques employed by the Japanese originated in
the United States, and had been elaborated by the Japanese into a management credo. Some U.S. companies, notably IBM and Hewlett-
Packard, have long been known for their effective quality control programs, and the Japanese agree that the best of the U.S. companies follow quality control procedures that are just as effective as their own.

After 1980, stiff Japanese competition was largely responsible for
making U.S. semiconductor producers more aware of the need for improved quality control procedures. Many producers realized that
quality could not be treated as a special parameter for which the customer had to pay extra; rather, it had become a central factor in the
competition for market share. Many U.S. companies established
quality control groups: Intel set up "Intel circles"; National Semiconductor established the position of vice-president for quality to drama-
tize its commitment to improving quality levels; AMD's chairman,
equaling the Japanese in the fervor of his exhortation to raise quality
levels, insisted that his company must be "second to none" in quality.

The case of AMD illustrates the kind of effort that U.S. semiconductor companies have made to improve quality. In early 1980, with reports of superior Japanese quality beginning to circulate, the company's management reached a decision that a major push had to be
made on quality. A three-step program was developed: (1) to determine through an internal study where AMD stood with respect to the
quality of its products on incoming inspection; (2) to dispatch company experts on quality control to domestic and overseas customers in
order to ascertain how they ranked AMD as a quality supplier and
what levels of quality they regarded as acceptable; and (3) based on
the results of the first two steps, to develop an international standard

of quality for AMD's products. Very few of AMD's customers could provide quantitative data, but a set of international standards was announced in January 1981.

As part of the AMD program, about twenty quality control circles were established, involving about 250 people or some 20 percent of the company's employees in the United States; quality control circles were also established at overseas plants. These circles were set up on a carefully planned schedule, with a built-in delay to allow for expansion only to the extent that well-trained facilitators were available to lead the quality control circles.

Long-range results have yet to be proved, but so far the circles have provided a good forum for involvement at a different level by a different group of people. Those who have taken part tend to show a heightened interest in their work and in the company. Some useful recommendations have come from those participating—for example, on how to get results of greater validity in temperature testing.

Besides instituting quality control circles, AMD undertook a number of other measures to raise quality. Test procedures were modified to incorporate additional sampling, and new tests were added prior to shipment to verify the results already compiled and to guarantee that "the quality that has been built in is actually there." Product flow was modified, and new tests were added to prevent subsequent problems. Checks were made to ensure that proper test procedures were followed and that calibration was accurate. AMD also expanded its program for the analysis of rejects and for feeding back that information to the area where the problem had occurred, so that corrective action could be taken to prevent future problems in those aspects of the manufacturing process. Since the company already had a long-standing procedure for very stringent inspection at early stages of production—such as the die level, prior to encapsulation—nothing further needed to be done in that area.

Some executives interviewed believed that the critical difference between many U.S. and Japanese companies on quality control lay not so much in the techniques employed as in the level of expertise and commitment of the workers involved. In their view, except for certain U.S. companies known for their low turnover rate, U.S. semiconductor firms would find it difficult to replicate the Japanese method for achieving quality control. In Japanese companies, it was said, there were experts at each station—epitaxy, masking, and so on—individuals with long experience on the job, who could spot flaws immediately and remove defective parts from the process. In the opinion of these executives, the fluidity of the labor force in the United States militates against such expertise and commitment. Typically, a U.S. engineer sitting in his office will review test data and detect flaws, but at a later stage than Japanese workers on the production line who can respond immediately. The Japanese approach, which saves money, may actually be closer to "early warning" than to prevention, but the key is clearly the special characteristics of the Japanese worker.

Clearly, the heart of the Japanese approach to quality control is the extraordinary stability and loyalty of the Japanese labor force and of Japanese management. Japan's lifetime employment system, even though it is not as widespread among workers in the semiconductor industry as many assume, ensures a relatively low rate of turnover and produces a strong sense of identification with the company. In the United States, particularly in California but to some extent in all areas, employees have been in the habit of moving from one company to another. The movement was slowed by the recession of the early 1980's, but in 1979 according to a survey conducted by the American Electronics Association, the rate of employee turnover in U.S. electronics companies was 45 percent for nonexempt employees, and smaller companies of 100 employees reached the astronomical figure of 78 percent per year. (For details, see Table 2.) The recession and the heightened possibility of layoffs caused turnover rates to decline dramatically in 1980. In San Francisco Bay Area companies, the turnover rate among nonexempt employees fell from 50 percent in 1979 to 37 percent in 1980. At National Semiconductor, the overall turnover rate dropped from 45.5 percent in 1979 to 19.5 percent in 1980. At Hewlett-Packard, however, which many have likened to Japanese companies in its personnel practices, the turnover rate remained constant from 1978 through 1980 at 8 to 12 percent.[42]

Even if the turnover rates when the economy improves do not return to the 1979 levels, the problem is still a serious one for the U.S. industry. As one manager noted, it is very difficult to maintain a reasonable degree of continuity on any project, and the high turnover rate among production personnel has very disruptive effects. The stability of employment in Japanese companies yields higher returns on an investment in training, which gives the company a greater incentive to offer expanded training opportunities to its employees. Training in Japan customarily includes special instruction in quality methodology. Moreover, since so many Japanese workers enjoy employment security, they are more likely to view automation as a means for enhancing the company's productivity, not as a threat to their livelihood. It is often said that in Japan, even labor favors automation.

TABLE 2

Employee Turnover Rates in the U.S. Electronics Industry in 1979

Category	Employees					Total
	1-100	101-250	251-500	501-1,000	Over 1,000	
All employees	59.10%	56.68%	50.17%	41.64%	27.15%	35.39%
No. of companies	280	175	102	89	89	735
Nonexempt	78.44%	72.40%	61.03%	49.24%	35.26%	44.70%
No. of companies	244	162	85	71	83	645
Exempt	28.03%	27.15%	24.22%	25.51%	15.33%	18.89%
No. of companies	213	161	85	71	83	613

SOURCE: American Electronics Association Benchmark Survey, 1980.

The Japanese claim that their workers not only have greater familiarity with the details of their jobs but are able to develop better working relationships with their associates, and they believe this makes for more effective communication within the firm, especially between labor and management. Contrary to the U.S. democratic ideology, there is traditionally much higher regard for bottom-to-top information flow in the Japanese factory than in a U.S. plant. Japanese executives assert that production workers play an important role in proposing qualitative improvements and, through the use of quality control circles, helping one another improve performance.

The Japanese semiconductor manufacturers also have the advantage of drawing their employees from a population base that is, broadly speaking, better prepared in mathematics and science. In math and science tests administered to ninth graders in the early 1980's, the Japanese outperformed their American counterparts. The overall achievement scores of Japanese engineering college graduates in math and science were the highest of any country in the non-Communist world.[43]

Some have suggested that Japanese culture may play a role. There is little doubt that the "group orientation" of Japanese society encourages cooperative behavior and organizational loyalty. Even the portion of the Japanese work force that is relatively transitory—such as young women who work on the assembly line an average of seven years and then leave to raise a family—tends to show extraordinary loyalty and commitment to the company. An executive at a major U.S. semiconductor producer noted that written Japanese requires a sensitivity to minor variations in the complex patterns found in Japanese characters, which could indicate a predisposition to perform well in jobs requiring an aptitude for pattern recognition. Finally, the virtue of cleanliness is far more important in Japan than it is in the United States, and in the semiconductor industry, where the maintenance of a dust-free environment is necessary for good quality control, habits of cleanliness are of especial value. Indeed, one Japanese executive remarked that the superior results attained by the Japanese in Hewlett-Packard's tests owed as much to the state of the Japanese companies' clean rooms as to any other single factor.

In considering these cultural influences, one must be very careful to avoid falling into a kind of cultural determinism. To say that certain aspects of Japanese culture may support goals of corporate loyalty or the maintenance of high standards of cleanliness is not to say that the Japanese are destined by virtue of their culture to be the leaders in quality control. On the contrary, for years the Japanese were known as producers of shoddy merchandise whereas Americans were believed to excel at quality. It may be that Japanese culture creates certain predispositions that are unusually supportive of efforts to establish exacting quality control standards in the semiconductor industry, but that by no means guarantees that the Japanese will be successful in doing so, or that those who operate in another cultural context will fail. Indeed, many Japanese are conscious of the possibility that social change in Japan could undermine the basic attitudes that have helped to bring about the success of their quality control efforts.

Automation. To what extent has Japanese production of memory chips benefited from an edge as a result of a higher degree of automation? The Japanese who were surveyed readily acknowledged that some major U.S. companies had production lines as fully automated as those of Japanese semiconductor manufacturers and in some cases possibly more so, but they firmly believed that the overall level of automation in the U.S. industry lagged behind that of Japan.

The Japanese recognized that a large percentage of the production equipment used by their semiconductor companies was imported from the United States, but they asserted that they had modified American-made machines to make them more efficient. For example, they had added attachments to raise the automation level of mask aligners, ion implanters, die-sorting machines, processing treatment machines, and final testing equipment. Toshiba engineers reported that U.S.-made automatic bonding machines, which operated at 0.3 seconds per die, had been modified to achieve a rate of 0.25 seconds per die. They emphasized that there had been numerous minor improvements of this kind in a great many areas. Even personnel on the production line had made suggestions—for example, modifications that would enhance the capacity of the machine to discriminate among wafers. U.S. semiconductor companies, it was suggested, did less modification, because there was a greater degree of interaction with the equipment manufacturer during the period when the machines were being designed.

Equipment developed in-house by the major Japanese semiconductor producers undoubtedly makes an important contribution to productivity and quality. The use of machines made in-house has the advantage of providing immediate feedback, which facilitates modification. Toshiba makes about 50 percent of its own machinery, and Hitachi an even higher percentage. High-speed automatic bonders, including computer-aided bonding systems, and plasma etching machines are examples of equipment that the Japanese have developed themselves. Recently, the Japanese have been making dramatic progress in automatic testing equipment. They still do not sell much in the United States, but between 1978 and 1981 they increased their share of the Japanese market from 30 percent to 60 or 70 percent.[44] One executive of a major U.S. semiconductor manufacturer predicted that the Japanese equipment makers were likely to provide extraordinarily formidable competition for U.S. manufacturers, but other U.S. executives questioned this assessment.

The Japanese edge in automated production facilities may be attributed to the earlier decision by U.S. semiconductor companies to make extensive use of cheap labor in assembly facilities located overseas in developing countries. Although this strategy depressed costs in the short run, it may have created some problems with respect to quality and productivity. A U.S. electronics executive seemed to lend support to that view when he asserted that there had been numerous problems with components that had been sent to Southeast Asia for die attachment and lead bonding. He contended that the lack of technical support people and engineers in the offshore plants made it difficult to maintain quality standards and to deal with unanticipated problems. The plants often failed to meet production deadlines or had problems in shipping that caused severe bottlenecks and upset the steady flow of shipments needed to keep production in the United States running smoothly. In some cases, parts shipped from Southeast Asian plants were even mislabeled.

The Japanese semiconductor industry has made limited use of overseas facilities. The Japanese answer to rising labor costs was mainly to modernize their assembly facilities. The automation of production lines not only served the original purpose of enhancing labor productivity but also helped to improve quality by providing a higher degree of cleanliness, uniformity, and reliability.

Not all U.S. executives agree that the Japanese are ahead in the automation of assembly lines. One executive emphasized that the United States had taken the lead in developing automated processes for the semiconductor industry, including IBM's C/4 process, Texas Instruments' automated front end line, Perkin Elmer's automatic alignment projection system, automatic lead bonders, automated testers, and automated wafer tracks. An American semiconductor equipment manufacturer felt that the Japanese had an edge in productivity because of the work habits of their labor force, not because of any higher degree of automation. Indeed, he said he had never seen a Japanese machine that had not been anticipated by prior developments in the United States.

Even if offshore assembly plants had initially been a disincentive to automation, this was no longer the case, a U.S. executive contended. Another American executive observed that equipment in offshore plants is completely replaced at approximately three-year intervals. Thus, he asserted, even if U.S. plants offshore once relied more heavily on labor, they are now highly automated. Some Japanese acknowledged that there was no difference between Japanese and U.S. technologies at the front end (e.g., wafer processing, doping, ion implantation, masking, etc.), but they insisted that the Japanese were ahead at the back end—particularly in the bonding process, which is the most time-consuming step. When it was pointed out that American plants in Malaysia had become more automated, as indicated by considerable increases in production without any increase in the number of employees, a Japanese executive responded: "Malaysian workers may have maintenance problems with our highest technology machines. If the Americans say they are fully automated in Malaysia, then it is a lower level of technology."

A spokesman for the U.S. industry took sharp issue with the allegation that the use of offshore facilities in Southeast Asia had created problems with respect to quality. He asserted that the sixty to seventy U.S. semiconductor plants in Southeast Asia had established an excellent record for productivity, quality, and low turnover. Asian process engineers were said to be excellent, with the best ones assigned to serve as plant managers, sometimes in other countries. Offshore assembly operations did have certain drawbacks, but sloppy work was not among them. The real drawbacks, he felt, were the difficulty of maintaining communication with decentralized and dispersed operations, the risk of hostile (or at least unhelpful) actions by host governments, and an unavoidable labor-intensive bias. He also noted that the U.S. companies had been taking their high technology operations—such as RAMs—offshore, whereas the Japanese were making increasing use of offshore facilities, but only for lower technology operations. U.S. technology in Southeast Asia, he asserted, was comparable to the technology that the Japanese employed at their plants in Japan. As further evidence of the U.S. industry's commitment to automation, a U.S. executive with long experience in the assembly equipment industry noted that U.S. companies increasingly were locating their new assembly facilities in the United States (and in Japan) rather than in low-wage developing countries.

The conflict between these two assessments is not easily resolved, but several conclusions seem warranted. First, although some U.S. companies pioneered the automation of the semiconductor industry, there seems little doubt that the Japanese made more extensive use of automatic lead bonding machines and have been doing so for a longer period of time. According to a variety of sources, the use of automatic lead bonding machines is virtually universal in Japan, whereas there are still some U.S. companies, particularly smaller ones, that rely on traditional bonding methods.

Even some senior executives of leading American semiconductor manufacturers agreed that the Japanese semiconductor industry practiced automation to a greater degree. One of these executives had originally argued vigorously that U.S. and Japanese levels of automation were comparable, but subsequently he reported that people in his company (Texas Instruments) who were "better informed and closer to the subject" had confirmed that over the five years between 1977 and 1981 the Japanese "practiced a broader degree of automation, particularly in assembly and test." He noted that Japanese companies were making more widespread application of pattern recognition; in addition, they not only had mechanized individual stations but had developed and applied transfer equipment between the stations. Hitachi's showcase 64K RAM production line is completely automated, including the inspection process, machines that scan for dust, and the transfer of materials; only ten people are employed on the line.*

Second, U.S. semiconductor companies are accelerating the pace of

* At the time of the visit to Hitachi by Weinstein and Linvill.

automation, but the process has proceeded more slowly because of the heavy U.S. investment in offshore assembly facilities. This is not to suggest that the decision to set up plants abroad was a mistake, but merely that there have been certain drawbacks associated with its cost advantages. Even the companies that are "fully automated" may not, in some cases, have reached Japan's level of automation. The Japanese assert that they are on the fourth or fifth generation of fully automated lead bonding equipment, whereas U.S. offshore assembly facilities may be on the first or second generation. An executive of a major U.S. semiconductor producer asserted that his company believed it was possible to produce about 600 units an hour per machine with automatic lead bonding equipment for 16K RAMs. He said he had heard of machines with astounding capabilities of up to 2,000 units an hour, but he had never seen such a machine and doubted that it could be built, given inherent limitations imposed by wire-feed capabilities. A Japanese executive reported that his company's machines produced on the average 900 units per hour, or 50 percent more than the number cited by the American executive. According to another U.S. industry source, NEC's new facility in Roseville, California, designed to produce 64K and subsequently 256K RAMs, will be a showcase of automation, substantially exceeding the current capabilities of the best of the U.S. semiconductor firms based in California.

One American executive acknowledged that the Japanese were more highly automated in several areas but questioned the significance of that edge. He confirmed that the Japanese made greater use of fully automated pattern recognition, whereas the U.S. companies generally used a semi-automated system, but he indicated that the difference in throughput between fully automated and semi-automated pattern recognition was probably less than 20 percent. There was, however, a substantial cost differential: fully automated pattern recognition machines cost $60,000 to $80,000 a machine; semi-automated machines cost about $20,000. Since U.S. companies were doing their assembly operations in low-wage countries, there was considerable saving in the semi-automated approach. There would, nevertheless, appear to be some advantage to the fully automated approach, since the same executive added that the U.S. companies were switching to fully automated pattern recognition now that it had become cost effective to do so. As for automated transfer stations, these were dismissed as "not high technology . . . just a mechanization thing," and a Japanese executive agreed with this assessment.

The same U.S. executive also acknowledged that Japanese die-attachment machines were the best available, and that the Japanese companies in general used better machines in this area than did their U.S. competitors. Japanese die-attachment machines were said to run at perhaps 2,500 units per hour. These machines, he said, were "definitely the way to go"; in fact, American companies were now making use of this equipment. But die-attachment machines, he emphasized, do not have the same impact on the production process as automatic lead bonding. They can help improve yield, but they are not likely to affect quality and reliability of the devices.

Is it likely that the U.S. companies can accelerate the pace of automation sufficiently to close whatever gap remains with the Japanese? There is nothing to prevent the U.S. companies from leapfrogging from first- or second-generation automatic lead bonding equipment to fourth- or fifth-generation machines, though it will not be easy. There may be political problems when accelerated automation means a net reduction in the number of jobs provided to local nationals in overseas locations. There may also be problems in training. It may be easier to close the automation gap by constructing new assembly facilities in the United States or Japan than by upgrading overseas plants. Moreover, Japanese production lines are also changing rapidly. The Japanese say that an LSI production facility lasts only about two years. The transition from the 16K RAM to the 64K RAM requires new equipment in certain areas, especially photolithography, but 16K RAM automatic lead bonding machines can be adapted for use in producing 64K RAMs by changing the program. One U.S. executive also made the point that differences between the U.S. and Japanese approaches to automation are partly attributable to the fact that Japanese firms can accept a lower return on capital, reflecting, among other things, greater tax incentives for savings and investment in Japan.

Quite clearly, there are some differences between the American and Japanese approaches to quality control and the degree of automation in the semiconductor industry, but the situation is changing rapidly and the future is hard to predict. Japanese and American executives alike agreed that whatever edge the Japanese may have gained in these areas is tenuous, and is essentially the result of management decisions that gave higher priority in Japan to improving quality and accelerating the pace of automation; it does not reflect any intrinsic differences between the two countries. To buttress this point about the critical importance of managerial priorities and leadership, Akio Morita, chairman of Sony, pointed out that Sony's San Diego plant has proved to be more efficient than his company's facilities in Japan.

The dramatic improvement shown by U.S. producers of memory chips after 1980-81 demonstrated even more clearly the impact that changing priorities could have on quality. According to Hewlett-Packard, there was a sharp improvement in the performance of U.S. suppliers of 16K RAMs in late 1980, and further improvement was recorded in a survey made in March 1981, when two of the three suppliers ranked highest in reliability were U.S. companies. Subsequent surveys indicated further progress by U.S. companies. Xerox reported a similar trend. Roger Dunn stated that U.S. suppliers to Xerox had made a "spectacular improvement" in quality after September 1980 and soon attained an average incoming-inspection failure rate in the 0.4-0.5 percent range. The improvement shown by U.S. firms may have been even more dramatic than these reports indicated, since these companies managed to narrow the quality gap with Japan at a time when the Japanese companies themselves were continuing to record significant improvement. In Xerox's view, as of late 1981, Japanese firms still held an edge, however. Xerox's greater confidence in certain Japanese companies was indicated by the fact that Xerox merely sampled the products of its two top suppliers, both of which were Japanese, whereas it tested all the products supplied by the U.S. companies and by the other Japanese firms.

It is important to remember that the claims of superior Japanese quality applied mainly to memory chips, not to semiconductor products generally. With respect to other LSI devices, such as microprocessors and peripheral chips, the U.S. industry was generally regarded as ahead. According to Xerox's Dunn, the Japanese were "coming on strong" with high-quality peripheral circuits, but they were "often one to two years behind the U.S. in LSI design." To illustrate the rough parity in LSI overall, Dunn reported that four out of fourteen (29 percent) Japanese LSI devices failed to pass qualification in 1980, compared with six out of twenty (30 percent) for U.S. makers. Xerox bought most of its microprocessors from Intel, whose product was said to be clearly the best; NEC, which second sourced the Intel line, was next in quality.[45] Thus, although certain Japanese companies were capable of posing a formidable challenge in any mass-produced semiconductor device, the great improvement in the quality of U.S. memory chips and the unexcelled quality of other U.S. semiconductor products strongly suggested that U.S. companies could equal, or surpass, Japanese quality when they decided that the *competitive* situation required their doing so.

Implications for Future U.S.-Japanese Competition

We have emphasized how difficult it is to make any projections for the future in the semiconductor industry, either in technological advances or in competitive strengths and weaknesses. One can reasonably assume that both the United States and Japan will continue to be strong in those areas of semiconductor technology in which they already exhibit special capabilities, and that both will continue to show impressive progress in overcoming what they presently recognize as weaknesses. But precisely how each industry will fare in relation to the other is impossible to say. Japan's weaknesses in software and new product development may be more far reaching than the widely publicized problems experienced by some U.S. companies with respect to the quality of their 16K RAMs, and they are long-term deficiencies. It seems likely that it will take the Japanese more time to overcome them than it will take the Americans to close the gap in quality control and automation. On the strength of its past preeminence and the breadth of its technological capabilities, the U.S. industry appears to be in a stronger technological position overall. The U.S. and Japanese engineers interviewed for this study tended to share that assessment, but there are three essential qualifications.

First, the ability of the U.S. industry to maintain this position of leadership depends on the availability of adequate resources—financial and human—and on the ability of management to see that the new emphasis on quality control and accelerated automation is maintained. Without denigrating the remarkable accomplishment of U.S. firms in improving the quality of their 16K RAMs, it should be pointed out that these improvements were recorded during a period of slackening demand in a recession. In a boom period it is more difficult to raise, or even maintain, quality levels, for pressures to increase production are intense and there are more inexperienced new employees. Second, the Japanese have already demonstrated their capability to take the technological lead, at least for a limited period of time, in selected areas on which they have chosen to concentrate substantial efforts.

Third, there is always the possibility that some new breakthrough could drastically alter the entire structure of the competition, much as the invention of the transistor affected companies dependent on the vacuum tube. For example, if the difficulties of working with gallium arsenide could be worked out and gallium arsenide came to be viewed as superior to silicon in optoelectronics and ultra-high-speed logic circuits, the Japanese lead in this revolutionary new technology could prove to be much more significant than the marginal increment to Japanese market share in RAMs about which so much concern is presently being expressed. That lead is, as we have said, perhaps a result of the fact that the U.S. companies chose to emphasize other approaches. Will that decision by the U.S. companies prove to have been wise?

The memory chip competition in perspective. The first major test, on which, as noted earlier, the press has already begun to focus attention, is the competition for the standard memory chip market, which is perhaps Japan's greatest area of strength. Japanese firms seem determined to marshal their forces to gain a large share of successive memory chip markets, and the challenge is indeed formidable.

Japanese corporations have some very important assets in this competition. Since the first 64K RAMs to reach the market in quantity were Japanese, this early edge enabled the Japanese suppliers to hold on to their market share by cutting prices faster than latecomers. Most of the factors that worked in favor of the Japanese industry with respect to the 64K RAMs will be even more important in the upcoming battle for the 256K RAM market, especially the availability of capital (needed particularly for RAM facilities) at comparatively low cost. As one U.S. executive put it, the RAM market will be dominated by whoever can lower the defect density and improve yield. Because the distance between individual components in newer generation RAMs is substantially less than that in earlier generation RAMs and the size of the memory cells is significantly smaller, the dust and alpha particle problems are much more severe.[46] Lowering defect density is in significant measure a function of cleanliness, an area in which the Japanese have always excelled.

But the U.S. industry has not conceded the memory chip field to the Japanese. Texas Instruments and Intel have produced designs that many regard as superior to Japanese designs, and though Japanese producers captured 70 percent of the 64K RAM market in 1981 and 1982, the U.S. market share grew significantly in 1983. Moreover, the American companies indicated their determination to avoid a repetition of the 16K RAM experience, when Japanese acquisition of a large share of the U.S. market was made easier by the inability of U.S. companies to meet requirements. That situation resulted from major (more than 50 percent) cutbacks in capital expenditures by U.S. companies during the recession of the mid-1970's. In the 1980-82 recession, the U.S. companies did not cut back so severely.[47]

Some industry analysts have already declared a Japanese "victory" in the standard memory chip competition. As of mid-1983 it was impossible to say whether the Japanese would end up with much more than 50 percent of the 64K RAM market. Who would dominate the 256K RAM market was even more uncertain. But let us assume for the moment that the Japanese do establish a dominant market position. What would be the implications?

Perhaps most important, a Japanese "victory" would not mean that U.S. producers would be driven out of the memory chip field. The difference between "victory" and "defeat" might be something like a 60 percent versus a 40 percent market share, or a 55 percent versus a 45 percent share. Benjamin Rosen, a respected financial analyst of the semiconductor industry, predicted that growth in demand during the 1980's would be so explosive that it would outstrip supply except during periods of recession. Other analysts, such as Thomas Kurlih of Merrill Lynch in New York, agreed.[48] But one senior U.S. executive, less optimistic, predicted that the industry would continue to follow its traditional pattern of fluctuating between peaks and valleys.*

It is necessary, moreover, to keep the memory chip business in perspective. RAMs have been described as the industry's "cash crop," needed to amortize investment in other products. The world market for integrated circuits in 1980 was about $8.7 billion, or about $10.7 billion if U.S. captive production is included; discrete devices accounted for an additional $3.7 billion worldwide. Annual sales of 16K RAMs amounted to $728 million, and the 64K RAM another $11 million.[49] The 64K RAM market in 1981 amounted to $100 million, but revenues from the sale of 16K RAMs dropped to $500 million as a result of falling prices, even though unit sales rose from 150 million to 240 million. According to Hambrecht and Quist technology analyst John J. Lazlo, Jr., dynamic RAMs accounted for only 7-8 percent of total semiconductor industry revenues. The Japanese scored an impressive achievement in capturing 70 percent of the early 64K RAM market, but remember that this represented only about 70 percent of a $100 million segment out of a $14-15 billion world semiconductor market.[50]

---

* There is a basic dynamic in the semiconductor industry that tends to exaggerate prevailing trends, thus creating a peak and valley effect. In a time of rising demand, production increases, and yields (the percentage of usable components) inevitably drop, as inexperienced new workers join the production line and pressures to produce make it hard to maintain quality standards. Falling yields lead to greater shortages and, ultimately, to rising prices. When demand falls, yields are pushed, producing more useable components and accentuating the condition of oversupply. This makes it possible to cut costs and prices.

Thus, RAM sales, important though they were, accounted for only a fraction of the overall semiconductor business. Moreover, RAMs were not the largest source of profit for companies that had a substantial memory chip business. The sale of RAMs represented no more than half the revenue for any of those companies, with the possible exception of Mostek. In addition, prices of the 16K and 64K RAM fell so sharply—for example, the 16K RAM fell from $4 to $2.50 in the last quarter of 1980 alone, and the 64K RAM fell from $28 in March 1980 to $5-6 in April 1982—that the view of the RAM as the industry's cash crop was called into question. Some analysts suggested that aggressive price cutting made the 16K RAM a "loser" that, far from being profitable, had to be subsidized by other semiconductor products. There would seem to be good reason therefore to question the assumption that the future strength of the U.S. semiconductor industry depended on capturing a majority of the 64K RAM market. To the extent that memory chips represent a "cash crop," the expanding market for those chips would make it possible for U.S. firms to draw enough revenue to amortize future product development, even if the Japanese end up with more than 50 percent of the 64K RAM market.

All the attention given to memory chips has rather obscured the importance of other semiconductor products, on which U.S. companies have a clear edge over their Japanese competitors. Estimates of Japan's hold on segments of the U.S. market have ranged from 18 percent (primarily memory chips) to as low as 5 percent overall.[51] Motorola's semiconductor chief, Gary Tooker, was quoted in early 1982 as saying that it would be "impossible for the Japanese to do in microprocessors what they've done in memory chips."[52]

From components to systems. The strong Japanese thrust in memory chips is, however, likely to speed a restructuring of the U.S. semiconductor industry that is already under way. According to an industry analyst who saw a Japanese victory in the 64K RAM competition as inevitable, the U.S. merchant semiconductor producers must compensate for their loss of market share in RAMs by producing not only components but also systems.[53]

The U.S. semiconductor industry has for some time been shifting its emphasis from components to systems. Nearly all the major merchant producers of semiconductor components have begun to place new emphasis on electronic systems. National Semiconductor now has a subsidiary that markets computer systems. Motorola purchased Four-Phase Systems, a computer manufacturer. Intel is much more heavily involved in software than it envisaged in the early 1970's and in the fall of 1980 opened a computer systems division. Intel indicated its determination to compete with the Japanese "on every front—in systems and in components." Systems accounted for about 40 percent of Intel's revenues in 1981, and this figure was expected to rise with new products in such areas as office automation, personal computing, and transaction processing.[54] At the same time, companies that built their reputations on systems were acquiring semiconductor companies with component manufacturing capabilities. Fairchild was absorbed by Schlumberger, Signetics by Philips, Mostek by United Technologies, and American Microsystems, Inc. by Gould; other acquisitions seem likely. Component producers are benefiting from the fact that major systems houses like IBM and AT&T are increasing their percentage of components purchased outside. (IBM, of course, has continued to account for a large part of the total world market in both spheres.) The reasons for this shift in emphasis from components to systems are essentially twofold.

First, the semiconductor component manufacturers have generally followed the strategy first suggested by the Boston Consulting Group to Texas Instruments—reducing prices rapidly as the company proceeds along the learning curve in the production of a new component. This strategy gives the company maximum market share and keeps it ahead of competitors, whose costs are always higher. The problem with this widely used strategy is that aggressive price cutting in order to gain market share for component sales results in perilously low profit margins, even when rapidly falling prices stimulate higher sales. Profit margins on electronic systems, on the other hand, have been high, and a systems manufacturer who builds his own custom semiconductor devices has a distinct competitive advantage over systems manufacturers who do not.

The move toward systems has also been spurred by the rapid development of computerlike structures centered around the microprocessor and its various implementations. In that sense, the trend toward systems may be said to be technology-driven.

Rise of the silicon foundry. Another counterbalance to a possible Japanese "victory" in the competition for the RAM market lies in the booming market for custom or semicustom chips, including not only captive but merchant production. A major new development of the early 1980's was the establishment of "silicon foundries," which produce custom semiconductor products for manufacturers, or even for other semiconductor companies. Intel, which previously had declined to become significantly involved in the custom chip business, built two special chips to be used in Ford automobiles and was said to be spending heavily to expand silicon foundry operations at its Arizona plant. Other companies, including National Semiconductor, did the same: AMD's president predicted that his company could succeed with only a small fraction of the 64K RAM market, partly by aiming high-priced chips at specialized segments of the market.[55] A number of companies were established solely to make customized or semicustomized chips. Wilfred Corrigan, who resigned as chairman of Fairchild intending to become a venture capitalist, decided instead to start his own customized chip company.

The increasing popularity of customized parts, which yield much higher profit margins than standard chips, led to some very optimistic predictions. Dataquest, for example, estimated in 1981 that custom and semicustom chips could account for as much as a third of the total chip business by 1985.[56] By 1983 custom chips already accounted for about one quarter of the total output of chips.[57] Although the boom in custom ICs anticipated in the early 1970's had failed to materialize, the situation had changed dramatically as a result of major progress in design automation. One industry leader said that design automation was "transforming the whole industry from mass-market, jelly bean producers to a value-added engineering-source business. Everything that made them successful in the past doesn't apply today."[58] The growing importance of customized chips seemed likely to strengthen the American companies vis-à-vis the Japanese.

Cross-investment. The competitive balance between the U.S. and Japanese semiconductor industries has become increasingly difficult to assess because of cooperative relationships between U.S. and Japanese companies and cross-investment, which tend to blur distinctions between the two countries' industries. A number of cooperative relationships had already come into being by 1983. For example, Hitachi was assisting Motorola in memory chip manufacturing technology in exchange for Motorola's design expertise and help in microprocessor technology. In one exchange, Hitachi described its high-performance C-MOS process to Motorola in return for a microprocessor mask set. Hitachi also agreed to supply Hewlett-Packard with 64K RAM masks, and indicated possible plans to sell production equipment and send engineers to advise H-P on 64K RAM production. Intel and NEC established a five-year cross-licensing, cross-compatibility, and technology exchange agreement, with NEC building Intel's microprocessor line and peripheral chips and Intel building some support products created by NEC, such as a double-density floppy-disk controller and a color graphics controller.[59] Amdahl and Fujitsu have worked together since 1976, and IBM and Matsushita have agreed to cooperate in the field of small computers. Other ties established included relationships between IPL Systems and Mitsubishi, National Advanced Systems and Hitachi, RCA Service Company and Hitachi America, Ltd., Sperry and Oki Univac Kashi, and TRW and Fujitsu.[60]

Even more significant was the sharp increase in investment by Japanese semiconductor producers in the United States and by U.S. semiconductor companies in Japan. American-owned plants in Japan served Japanese and other Asian markets, which were expected to grow at above-average rates. Texas Instruments, which has maintained a manufacturing facility in Japan for some time, is producing 64K RAMs in Japan. Intel was scheduled to produce 64K RAMs and microprocessors in Japan beginning in 1983, and IBM may also build RAMs in Japan. Motorola is already producing ICs in Japan through a joint venture undertaken in 1980 and is expected to build a new plant to make 64K RAMs. Other U.S. companies, including Fairchild, are also said to be preparing to build factories in Japan, and some, including National Semiconductor and Intel, have established design centers there.[61]

As for the Japanese-to-America move, NEC's showcase plant in Roseville, California, which will produce 64K RAMs and eventually 256K RAMs, as well as custom chips, will nearly double direct investment by Japanese semiconductor producers in the United States. Toshiba has a large manufacturing facility in Sunnyvale, California, where it produces 16K RAMs and custom-made logic ICs and other products. Hitachi tests and assembles ICs at a Texas plant it bought in 1978. Fujitsu is setting up a design center in Santa Clara, California, to support its gate array business in the United States, and it plans a metalization facility in San Diego, where it maintains test and assembly operations. Sony has for some time had a technology bank in Palo Alto, staffed by an international group of fifteen engineers; this is an independent R&D facility that gives Sony some of the advantages of a U.S. approach to research.[62] One effect of the growth of Japanese investment in the United States has been to mitigate somewhat the communication problems that handicapped the Japanese in the custom IC design business. As noted above, the new NEC and Toshiba facilities were explicitly aimed at strengthening the ability of those companies to compete in the custom chip market.

The rush of cross-investment complicates the task of compiling trade statistics. For example, when Japanese exports of 16K RAMs to the United States declined, this was believed to be due not only to falling prices but also to the increase in production of 16K RAMs by Japanese factories in the United States. Increasing exports of uncased wafers from Japan to the United States were one indication of this.

More important, however, is the impact of this cross-investment on the competitive position of the two countries' industries. By producing 64K RAMs in Japan, the U.S. companies hoped to gain access to the Japanese market, take advantage of Japanese manufacturing efficiency to produce devices for export to other markets, and gain access to Japanese capital. By establishing factories in the United States, Japanese companies hoped to gain better access to the U.S. market, including the custom and semicustom markets, and at the same time gain access to U.S. software-engineering talent. One industry expert predicted that by the mid-1980's as much as 50 percent of each country's VLSI products might be produced in the other, and each country's industry would be able to draw on the other country's strengths to overcome its own weaknesses.

Thus the danger of either the U.S. or the Japanese semiconductor industry's overwhelming the competition seems remote. Although particular companies may come under strong pressure, it would be a serious mistake to view Japanese success in the memory chip business as an indication that the U.S. semiconductor industry is about ready to fall before the sort of irresistible Japanese onslaught that decimated a series of more "mature" U.S. industries, including the automobile industry. Silicon Valley is not Detroit, and there are striking differences between the semiconductor industry and other fields in which the Japanese have gained market share.

First, the technological component is much higher in the semiconductor industry. Given the extraordinary pace of technological advance in the semiconductor industry, the technological edge has greater commercial relevance here than in many other industries. It is unusually rewarding, from a commercial standpoint, to be first with new products, and this happens to be the U.S. industry's strong suit. Second, unlike U.S. industries in which declining productivity has been a problem, the American semiconductor industry has established a record of consistently high increases in productivity. Third, the semiconductor industry is a growth industry, with enormous prospects for the development of new markets. Fourth, there is early recognition on the part of the U.S. industry of its shortcomings. Even though Japanese memory chip producers may have a certain edge in quality and automation, it is nothing like the gap that exists in the automobile industry. In the semiconductor industry, measures to correct known shortcomings are already bearing fruit. Finally, in contrast to steel or automobiles, the semiconductor industry in the United States has a strong and resilient entrepreneurial element, which allows it to respond more freely and quickly to challenges than is possible in most of the older industries.

Will critical resources be available? In sum, competition between the U.S. and Japanese semiconductor industries is certain to intensify over the next decade. But recalling Rosen's prediction, the growth of the semiconductor industry during the remainder of the 1980's is likely to be so dramatic that there need be no loser in that competition. The question is whether the resources—both financial and human—required to remain competitive will be available. The requirements for these resources will increase dramatically during the coming decade. They are almost certain to be made available to the Japanese industry, for reasons that will become clear in the subsequent chapters of this study. Prospects for the U.S. industry in this respect are less certain. If it does not have the essential resources, the U.S. semiconductor industry will be unable to maintain the technological edge that is the necessary condition for maintaining lively competition with the Japanese.

FOUR

Political Context

Daniel I. Okimoto

The Pioneer Industry: Birth and Early Growth in America

The evolution of the semiconductor industry suggests that the need for, and relative effectiveness of, direct and indirect government measures vary with time and circumstances. Typically, for high technology industries, it is during the formative stages of growth that direct assistance is most needed and registers the greatest impact. That was certainly the case in the United States and Japan where the two governments, seeing the need and seizing the opportunity, moved decisively to create and nurture the semiconductor industry. Both Tokyo and Washington relied heavily on funding for research and development and on guaranteed procurements to get the enterprise launched. Both governments worked closely with industry in seeking to link national interests with end-product applications technologically within reach. As a result of these direct measures and of efforts mounted in the private sector, the semiconductor industries in America and Japan got under way earlier and grew more rapidly than anywhere else in the world.

Once on its feet, however, the semiconductor industry quickly "outgrew" its need for, and dependence on, government help. The early military orientation in the U.S. gave way to purely commercial incentives as the industry's basic driving force propelling it toward maturity. The industry thus changed in ways that diminished the effectiveness of a direct government role. At the industry's present stage of development, the principal effect of state policies has to do with its interactive impact on the private sector, specifically in terms of whether it enhances or retards market competitiveness.

Military-related research and development. It is no accident that the semiconductor industry was conceived, born, and nurtured in the

United States. Not only did the U.S. have the requisite economic and technological infrastructure; but as the world's military hegemon, it also had the incentive to push state-of-the-art technology in directions that promised big payoffs for national security. This blend of factors has made America the postwar cradle for a number of high technology industries, including atomic energy, space and aeronautics, information processing, and telecommunications. The government's role in the early development of these industries was central.

In nearly all these industries, the undifferentiated nature of military and civilian technology meant that government-backed military "technology push" would converge with, and accelerate, incipient commercial forces, and that this would eventually yield bountiful "spillover benefits" for the civilian economy. Over time, huge corporate clusters sprang up around these technologies as jet airliners, computer hardware, consumer electronics, and semiconductor devices began to be sold in large quantities on the marketplace. A number of American corporations succeeded in making the transition from serving as strictly military vendors to selling predominantly in the commercial market; other firms, including many new entrants, entered the field after military demand had peaked. With little or no foreign competition, and with the lion's share of world demand concentrated at home, America's corporate pioneers had no difficulty establishing positions of pre-eminence in technology, production, and sales.

The early development of the semiconductor industry is closely associated with government sponsorship of research and development (R&D) and military and space-related procurements. Semiconductor research, which began in the prewar period and continued through the Second World War, was stepped up in the postwar era. Bell Laboratories, the huge research arm of American Telegraph and Telephone (AT&T), played a leading role in the early phases of postwar R&D. It served as a kind of "seedbed" for basic research and innovations, including the invention of the transistor in 1948, and provided a training ground for key engineers and scientists who went on to make major contributions working for newly formed companies in the industry. The U.S. military became very interested in the potential application of semiconductor technology for space vehicles and aircraft, believing that electronic miniaturization could upgrade their performance characteristics. The Air Force, and then the Army and the National Aeronautics and Space Administration (NASA) underwrote a significant portion of early R&D. In 1957, for example, about 70 percent ($518 million) of all R&D for electronics and communications equipment came from government sources.[1] Even as late as 1968, the figure still hovered around the 60 percent level ($1.5 billion).

The figures are substantially smaller but still significant if we look only at semiconductors; in 1958, the government accounted for about one quarter of total semiconductor R&D, and if indirect funding is included, the sums were somewhat higher (see Table 3).[2]

Coming at a time of greatest need (i.e., the early phases of development), government support helped to advance R&D in ways that went beyond dollars and cents.[3] It represented, in effect, a sanctioning of the whole attempt to carve out a new field of microelectronics. It communicated to private enterprise the message that, if microelectronic products could be developed, military and space programs stood ready to utilize them. The sense of urgent need gave companies greater incentive to invest in semiconductor R&D, and government sponsorship also reduced the perceived risks of chartering new technological frontiers by altering corporate calculations of the risks-to-returns ratio. Development time, as calculated from the start of exploratory research to the production of new devices, was also accelerated. High reliability standards were set by the stringent needs of military and space applications. And the high priority given to R&D related to national security prompted the federal government to support basic research at some of the leading universities, including Stanford, M.I.T., and Pennsylvania, creating a trio of actors (government, industry, academia) that dynamically pushed forward state-of-the-art technology.

The government, therefore, played a constructive role in terms of providing a "technology push" for the early development of the semiconductor industry. Foreseeing the potential value of microelectronics, the military prodded industry to focus efforts on miniaturization, and provided a significant portion of the seed money to carry out R&D. Well-established firms with experience and existing facilities and scientific personnel, such as Western Electric and the major receiving-tube firms (General Electric, RCA, Sylvania, Raytheon, and others) were awarded a large share of the early government grants.[4] It should come as no surprise, accordingly, that the Bell Laboratories and the receiving-tube firms pioneered much of the basic technology during this period. Other firms broke into the circle somewhat later, winning defense and space contracts in the late fifties and early sixties. Texas Instruments (TI) won a $1.15 million contract from the Air Force in 1959 to develop integrated circuits; less than a year later it received a $2.1 million contract for the development of equipment to fabricate

TABLE 3

U.S. Semiconductor R&D Expenditures, 1955-1961

(Billions of dollars)

Use of funds	1955	1956	1957	1958	1959	1960	1961
R&D	3.2	4.1	3.8	4.0	6.3	6.8	11.0
Production refinement:							
Transistors	2.7	14.0	0.0	1.9	1.0	0.0	1.7
Diodes and rectifiers	2.2	0.8	0.5	0.2	0.0	1.1	0.9
TOTAL	8.1	18.9	4.3	6.1	7.3	7.9	13.6

SOURCE: John E. Tilton, *International Diffusion of Technology: The Use of Semiconductors* (Washington, D.C.: Brookings Institution, 1971), p. 93.

integrated circuits in large quantities. Motorola received federal support for integrated circuits for communications, as did Fairchild for micrologic circuits, RCA for metal oxide semiconductors (MOS), and Westinghouse for miniaturization.

Government backing permitted firms directly and indirectly to defray a variety of costs: engineering design and development, production process equipment and technology, testing facilities, and research and administrative personnel. Federal support also contributed to the financial well-being of some firms in that R&D expenditures, particularly for high-risk ventures, did not have to come entirely out of current sales and profits, which meant that retained earnings could be used for new investments or other purposes.

The U.S. system of national R&D, centered around military security, has lent itself to the pioneering of high technology industries. Thanks to the overlapping of civilian and military technology, the government has been able to support sophisticated, state-of-the-art R&D for military applications that have had substantial spin-off benefits for the civilian economy. For costly, exploratory, uncertain, and therefore highly risky undertakings involving national security, the state has demonstrated a willingness to step in to cushion the costs and absorb some of the risks.

Of course, such support has yielded high returns only to the extent that it has succeeded in drawing upon, reinforcing, and complementing R&D efforts in the private domain. In this sense, the role of state sponsorship must not be exaggerated; it ought to be characterized as supportive and supplemental, not primary. The main impetus for technological advancement has always come from the private sector, even during the early stages. To put the government's role in proper perspective, it should be pointed out that, for all the millions poured into R&D, few decisive breakthroughs in semiconductor technology can be attributed directly to federally funded projects. Nearly all the basic technology—the transistor, planar process, microprocessor—was developed in the laboratories of private industry in projects that were internally funded.

It would be wrong also to assume that government-sponsored R&D came costfree. As federal funds veered away from basic to applied research, the narrowing of focus began to exact opportunity costs. At the firm level, as well as in the aggregate, hard questions had to be asked: Is participation in applied research for military objectives and devices the best use of research personnel, time, and facilities? Does it divert limited manpower from commercially more attractive projects? Can companies operate on a stable, long-term basis, if they are dependent on military contracts? Or are profit and employment levels bound to feel the destabilizing effects of fluctuations in government support? Defense contracting bred inefficiency because firms bid and operated on a cost-plus basis.[5] Being concerned with performance and specifications, the system failed to structure incentives in ways that forced firms to be as cost conscious as they would have been operating only in the commercial marketplace.

The nature of basic research is such that outcomes are almost always unclear. This implies that the extent to which the market mechanism can be counted on to sustain basic research is limited; hence the need for government to step in and fill the breach. The government is not particularly adept at product development and commercial application; once exploratory research is brought to that stage, the structure of incentives and the ratio of returns to risks are usually favorable enough that private firms can be expected to compete to be the first to market. Such market-related dynamics are reflected, however crudely, in the negligible number of semiconductor patents held by the U.S. government: from 1969 to 1979, only 4.7 percent of all semiconductor-related patents and 2.1 percent for device patents (1968-80). As suggested in these figures, product application is most efficiently left in the hands of private enterprise.

Where the government can reinforce market dynamics is not simply in the timing and level of financing, but also in the choice of which firms will receive R&D contracts. Mention has been made already of the fact that the bulk of early government contracts (78 percent in

1959) went to such established companies as Western Electric and eight large producers of receiving tubes. As technology advanced, other companies, some of them of the new start-up variety, began winning a larger share of contracts. Eventually, the technological vitality of these companies came to be reflected in the growing number of patents registered under their names.[6] The U.S. government's pattern of R&D sponsorship facilitated the diffusion of semiconductor technology. In combination with procurement policies, it helped to keep the barriers to new entry low. The highly diversified structure of firms, which government R&D, procurement, and antitrust policies helped to shape, has been a major source of the U.S. semiconductor industry's dynamism and strength.

According to one estimate, current government expenditures account for only about 5 percent of all semiconductor R&D.[7] Private companies are plowing back a substantial portion of profits into research and development, since many companies must remain at the cutting edge of technology in order to stay competitive, and the rewards of bringing new product lines to market can be substantial. Clearly, the U.S. semiconductor industry has long been at a stage where market "demand pull" is the principal driving force behind technological innovation. On the other hand, the U.S. government is probably still the only source of large funding for basic, exploratory research requiring huge capital outlays on a very high risk basis. It alone is able to assume such risks based upon the legitimation of advancing national security. If the semiconductor industry's commercial position is enhanced by defense-related research, it is clearly a by-product of research designed primarily to achieve military objectives.

Government procurements. An effective government strategy for the promotion of high technology industries (extrapolating from the experience of semiconductors) is one that pursues a two-tier approach, combining "technology push" with "demand pull."[8] Sensing that a basic technological breakthrough is possible, or even close at hand, is normally not enough to make it happen. A powerful market-type incentive is needed as a "forcing mechanism" to push technological open-endedness across the product threshold. Knowing that there is an urgent need and a guaranteed demand for new products can serve as that kind of incentive. It lowers the perception of risk while boosting the projection of benefits. The impact of public procurements is particularly significant during the early phases of development, as companies are struggling to make the transition from products on the

TABLE 4

U.S. Production (or Shipments) of Semiconductors
for Defense Consumption, 1955-1977

Year	Total (million $)	Defense (million $)	Defense as percent of total
1955	40	15	38%
1956	90	32	36
1957	151	54	36
1958	210	81	39
1959	396	180	45
1960	542	258	39
1961	565	222	39
1962	571	219	38
1963	594 (600)	196 (213)	33 (36)
1964	635	157	25
1965	805 (879)	190 (194)	24 (22)
1966	975 (1,055)	219 (254)	22 (24)
1967	879 (1,074)	205 (297)	23 (28)
1968	847 (1,189)	179 (274)	21 (23)
1969	1,457	247	17
1970	1,337	275	21
1971	1,519	193	13
1972	1,912	228	12
1973	3,458	201	6
1974	3,916	344	9
1975	3,001	239	8
1976	4,968	480	10
1977	4,583	536	12

SOURCE: Robert W. Wilson, Peter K. Ashton, and Thomas P. Egan, *Innovation, Competition, and Government Policy in the Semiconductor Industry* (Lexington, Mass.: Lexington Books, 1980), p. 146.

NOTE: Figures in parentheses for 1963 and 1965-68 are shipment rather than production data; figures for 1969-77 are rounded.

drawing boards to commercially marketable goods. Through procurements the government can serve, in effect, as the first customer and end-user.

As seen in Table 4, defense consumption accounted for well over one-third of total U.S. production from 1955 to 1963, reaching a high of 45 percent in 1958. Looking only at integrated circuits (ICs), we see that the government in 1962 bought up to 100 percent of total IC production (see Table 5). Semiconductors were used extensively for installation in Minuteman missiles and the Apollo spacecraft. Around 1964, with commercial opportunities expanding rapidly, the military proportion of total production began falling off. By 1977, defense-

TABLE 5

Government Procurement of Integrated Circuits
Production, 1962-1968

Year	Total production (*million $*)	Defense procurements as percent of total
1962	4	100%
1963	16	94
1964	41	85
1965	79	72
1966	148	53
1967	228	43
1968	312	37

SOURCE: John E. Tilton, *International Diffusion of Technology: The Case of Semiconductors* (Washington, D.C.: Brookings Institution, 1971), p. 91.

related consumption constituted only 12 percent of total production; but in absolute dollars, the amount continued to rise steadily. Defense procurements came to $536 million in 1977, nearly ten times the figure twenty years before.

Military procurements were instrumental in stabilizing demand during the industry's toddler stages. This functioned to offset the inevitable problems associated with being first, such as very steep start-up costs, initial uncertainties concerning output, and some degree of trial and error in commercialization. The pioneering companies did not have the advantages of latecomers, such as lessons learned from the first experience, existing markets, and the availability of extant patent rights,[9] and there are many examples of first-to-market companies (like Univac, maker of the world's first computer) that never succeeded in working out the early problems. Often second or late-to-market companies make the product improvements and devise the marketing strategies necessary to exploit the commercial opportunities created by technological breakthroughs. In semiconductors, the period of first-to-market monopoly before competitors come out with the same product has contracted. Pioneer firms, which could look forward to a period of several years of monopoly on new products, now can anticipate a much shorter period—for some products only a few months—before competitors introduce similar products (see Fig. 2). The same trend is discernible with respect to plant facilities. What used to last six, seven, or more years must be replaced more quickly, if the firm is to stay competitive (see Fig. 3). Contracting product and equipment life cycles not only raise the level of capital requirements but also limit

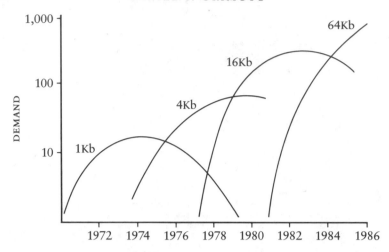

Fig. 2. Storage capacity and demand for MOS RAM, 1970-1986. (Source: *IC Shuseki Kairo Guidebook, 1981* (Tokyo: Nihon Denshi Kikai Kogyokai, 1981), p. 27.)

the time within which "up-front" R&D and plant investments can be recouped. If life cycles contract further, incentives to innovate might be seriously dampened. Thus there are definite costs and risks (along with benefits) associated with being first to market.

Since new product development is usually expensive, with early batches produced only in small quantities, per unit costs tend to be high. In new markets it is very difficult to recoup costly investments unless there is some way of expanding volume or of subsidizing prices. Military procurements provided both means. Figure 4 shows that, during the early years, the average price of transistors sold to the DOD was several times higher than the price of those sold on the open marketplace. This functioned, in effect, as a kind of hidden subsidy. Companies were able to set commercial prices lower than strict calculations of costs and profits would have allowed, because the comfortable margins obtained through military sales could be used to make up the difference. Military procurements thus provided a solid foundation for fledgling companies seeking to establish a base for commercial operations.

Guaranteed sales would have strengthened the semiconductor industry, even without such glaring price disparities. Semiconductor production reflects the dynamics of learning curve theory, which postulates that costs fall in direct proportion to increasing volume of pro-

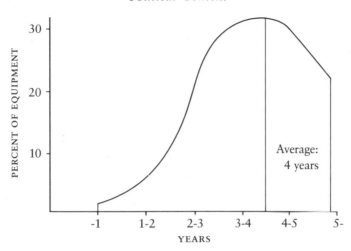

Fig. 3. The lifespan of Japan's IC equipment, 1981. (Source: *IC Shuseki Kairo Guidebook, 1981* (Tokyo: Nihon Denshi Kikai Kogyokai, 1981), p. 33.)

duction. This means that larger volumes and longer production runs lower per unit costs and move companies down the learning curve.[10]

The combination of defense-related R&D and military procurements also may have enhanced the semiconductor industry's capacity to secure capital. We have already explained how the fungibility of federal support could be converted into commercial assets. By signaling the government's confidence in, and commitment to, the industry's development, public procurements and R&D support also helped to instill the public confidence necessary to secure financial backing in the capital market—in stocks and bonds, bank credit and loans, and venture capital. To the extent that government backing enhanced corporate capacity to raise capital, it contributed directly to the semiconductor industry's commercial consolidation and capacity to outgrow its early dependence on the military.

The semiconductor industry's early reliance on the military had certain drawbacks of course. Though it advanced the industry down the learning curve, it may have also retarded responsiveness to commercial opportunities, and this problem has been compounded by the inadequacy of institutional mechanisms for industry-specific analysis and policy coordination. Washington has no counterpart to Japan's Ministry of International Trade and Industry (MITI), which can coordinate technological potential with industrial and economic needs. As

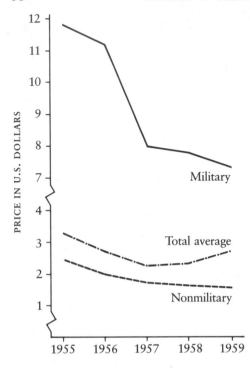

Fig. 4. Average transistor prices in the U.S., 1955-1959. (Source: H. Kleinman, *The Integrated Circuit* (Ann Arbor, Mich.: University Microfilms, Inc., 1966), p. 81.)

Robert Gilpin points out, the American government, lacking microeconomic planning to go with macroeconomic management, has not had notable success in such coordination.[11]

The country's tardiness in exploiting technological opportunities for commercial gain is seen in the field of solid state physics. Although American scientists did pioneering theoretical work in solid state physics, as well as in its technological applications, most of the applied work was oriented toward potential military and industrial end-uses. Japanese scientists and engineers, on the other hand, building upon the work done in America, concentrated on consumer goods and soon established themselves as world leaders in mass-produced, consumer durable goods like transistor radios, television sets, stereophonic equipment, and video tape recorders.[12] The United States failed to take advantage of its early technological and scientific lead and the decisive assets at its disposal.

The field of consumer electronics is now an immense industry in Japan, accounting for a significant fraction of its merchandise exports as well as its domestic semiconductor consumption. In 1981, the consumer electronics industry took about 50 percent ($2.2 billion) of Japan's total production of semiconductors ($4.4 billion), a figure far in excess of the U.S., where consumer goods accounted for only around 15-20 percent.[13] In the United States, production is geared more to military, industrial, and electrical machinery use, especially computers. Such differences in market orientation reflect conscious choices made in the past. With the U.S. concentrating on military and industrial applications, and Japan lagging behind in state-of-the-art technology, Japanese companies had little choice but to find niches in the consumer electronics market; the sale of handheld calculators in the sixties gave the demand for semiconductor components a particularly strong boost.

The present composition of end-user demand is the outgrowth of two relatively distinct patterns of development. In the United States, the long-term legacy left by the federal government, which inherited the mantle of world hegemon, is apparent not only in the nature of the early development of the semiconductor industry but also in the kind of support-system industries that grew out of it—producers of military hardware, aerospace companies, and computer manufacturers, all of which have received nurturing from military procurements and R&D. The semiconductor industry would not have grown so quickly if the government had not played so central a role in its initial development (and that of related industries). Here again, we are drawn back to two earlier hypotheses: (1) that America's system has been admirably suited for the creation of new high technology industries where military and commercial technologies have been undifferentiated; and (2) that the same system may not be as efficiently designed to compete with countries singlemindedly oriented toward commercial applications, owing to divergent military and commercial trajectories, and the "crowding out" effect of military programs. By comparison, Japan's system has been fine-tuned to pursue a second-to-market or follower's strategy, based on mass volume production for consumer goods. Each system has strengths and weaknesses.

So far, Japan's strategy, dictated by the imperatives of technological constraints and of comparative advantage, has been to concentrate on selected products at the lower end of the market spectrum. Such standardized products as random access memory chips could be mass produced and sold aggressively worldwide, taking advantage of Japan's manufacturing and marketing prowess. By starting at the lower end of the spectrum, Japanese manufacturers could lay the groundwork to climb their way gradually up the ladder of higher value added. For some U.S. merchant houses, Japanese inroads into standardized, mass-volume product markets have posed a serious threat. The logic of semiconductor technology and commercial competition is such that leading first-comer firms cannot afford to concede mass-volume markets to foreign competitors, and remain content simply to concentrate on higher value added segments. This permits "pursuers" to carve out and consolidate market footholds, and denies "pioneers" an indispensable technological, manufacturing, and marketing base. Eventually, without that base, the next generation of memory chips (256K RAM) and higher value added markets will be harder to hold. The "low-to-high" technology and value-added approach is one that the Japanese have successfully followed in displacing American front-runners in a variety of other industries.

On the other hand, state-of-the-art technology in semiconductors is enormously complicated, and making the transition from standardized devices to highly sophisticated products and system designs is not easy. Japan has so far made inroads mainly in very narrow segments of the market, such as MOS memory chips and CMOS (complementary metal-oxide-silicon) static random access memories (see Table 6). It has not advanced very far in other sectors of the U.S. market such as discrete, linear, digital bipolar devices, CMOS logic, and custom circuits and microprocessors; nor, over the near term, is it likely to capture a large U.S. market share in many of these areas. In view of the imposing technological hurdles still ahead, and America's commanding lead in most product markets, the "threat" of Japanese displacement through a strategy of "low-to-high" appears either exaggerated or premature.

Nevertheless, Japan's offensive is proving difficult for a number of U.S. merchant houses to deal with. Price competition in 64K RAMs, for example, is so cutthroat that many of the U.S. companies that manage to stay in certain product markets are finding the profits earned meager at best. Price cutting is especially difficult for small merchant houses because large vertically integrated Japanese firms can take advantage of captive sales and volume production to price aggressively.* What makes the situation so frustrating is the seemingly asymmetric openness of the American market, compared to the diffi-

---

*Some American firms have accused the Japanese of outright dumping. Motorola has been among the most vocal critics in this regard. W. J. Sanders, chairman of the Board of Advanced Micro Devices, took out a full-page ad in the *Wall Street Journal* in which he said, "I just don't want to pretend I'm in a fair fight. I'm not . . . Are they [the Japanese] dumping? I think so. It sure looks like dumping." *Wall Street Journal*, Apr. 5, 1982.

TABLE 6

Japanese Production of Integrated Circuits, 1978-1981

(Y100 million)

Type of circuit	1978	1979	1980	1981 est.	Percent change, 1978-81	Market share, 1980
MOS memory	457	840	1,388	1,503	48.7%	27%
MOS logic	860	1,123	1,710	2,563	43.9	33
Bipolar digital	409	511	726	863	28.3	14
Bipolar linear	791	958	1,333	1,628	27.2	26
TOTAL	2,517	3,432	5,157	6,557	37.6%	100%

SOURCE: *IC Shuseki Kairo Guidebook, 1981* (Tokyo: Nihon Denshi Kikai Kogyokai, 1981), p. 18.

culty of penetrating the Japanese market. Of various barriers to entry, there are such "unintentional" obstacles as formidable language difficulties, unfamiliar business practices, fierce product competition, complex distribution problems, and hard-to-break-into industrial groupings (*keiretsu*), which do not constitute "unfair" trade practices under GATT (General Agreement on Tariffs and Trade). But many American executives believe there are also "deliberate" or "intentional" barriers that are thrown up with the specific aim of keeping American competitors out: a tacit and widespread inclination to "buy Japanese," a willingness to buy only sophisticated products that the Japanese do not produce themselves, a past history of infant industry protection, and so on.

Which constitutes the greater obstruction—unintentional or deliberate barriers—is a matter of debate. The Japanese attribute America's problems to a lack of proper preparatory work, persistence, and effort; the Americans contend that greater weight should be placed on what they see as a hard-shell, thoroughly nationalistic system that keeps them locked out. Japanese exports to the U.S. in 1982 exceeded $600 million whereas imports from the U.S. and Europe came to slightly over $400 million, leading to an imbalance of over $200 million.[14] The year before that, the U.S. enjoyed a trade surplus of $20 million. This suggests that trade flows fluctuate each year, sometimes substantially, even though the trend seems to be moving decidedly in the direction of larger Japanese trade surpluses. Japan's share of the total U.S. semiconductor market came to about 10 percent in 1982, with MOS memory chips constituting the largest category by far, 25 percent of that market; Japanese imports accounted for less than 7 percent in other product categories. Trade is thus not very large relative to domestic sales (not including captive supplies). U.S. industry executives feel they ought to be exporting much more to Japan, given the U.S. industry's comparative strength and trade performance in other parts of the world.

American industry leaders realize that Japan cannot be allowed to monopolize its own market and use that as a base to expand its share of the U.S. market. If the experience of the automobile industry is an indication, such a pattern would be a formula for long-term defeat, particularly in an industry that is as sensitive to the effects of learning curve as semiconductors. Japan's rise as challenger increases the importance of establishing a U.S. sales and a manufacturing presence in Japan. That market is simply too big and vital to ignore; but in spite of its recognized importance, not many American firms have taken the bold step of setting up manufacturing facilities in Japan.[15]

The Japanese "threat." The semiconductor industry is perhaps the first instance where the Japanese are mounting a sustained long-term challenge in the field of high technology, America's traditional preserve of comparative advantage. The U.S. semiconductor industry, determined not to suffer the same fate as some of the old-line manufacturing concerns, has responded to the Japanese "threat" by amplifying and using it as a lever to apply pressure on both the Japanese and U.S. governments. The U.S. industry wants Japan to eliminate what it considers "unfair" trade practices, and it wants its own government to establish a better business environment, including better supply-side incentives to keep it growing and strong.

In seeking to eliminate "unfair" trade practices, America's semiconductor industry has been able to ride the swelling tide of resentment over the large U.S.-Japan imbalance in merchandise trade.[16] The U.S. government has been willing to bring pressures to bear, particularly during periods of high unemployment, in order to force the Japanese government to agree to the removal of nontariff barriers. After protracted negotiations, for example, NTT (Nippon Telegraph and Telephone) reluctantly agreed to open procurements to foreign bidding. The Semiconductor Industry Association (SIA) also worked hard to introduce reciprocity legislation on high technology, one of many reciprocity bills sent before the Congress.* (The USTR (United States Trade Representative) Office has wanted to have authority to negotiate on reciprocity over high technology and the services.) As the reciproc-

---

* As of September 1982, over twenty bills aimed at achieving reciprocity in foreign trade had been introduced in the 97th Congress. I am indebted to Senator Thomas Eagleton (D-Missouri) for explaining the situation, Sept. 1, 1982.

ity bills now stand, the emphasis is no longer "substantially equivalent competitive opportunities," as originally drafted, but "fair and equitable market opportunities," meaning, in effect, that the U.S. government will not tolerate discrimination against American firms operating in Japan; it wants them to be free to participate in national research projects, just like Japanese firms, and to apply for public and private financing.

Anticipating such demands, the Japanese government has invited American firms to participate in such projects as the Fifth-Generation Computer Project; so far, the American response has been lukewarm. Working out the details for joint research is never easy, and it remains to be seen to what extent, if at all, cooperation can be realized. The Japanese government is also considering a foreign application from Fairchild Camera and Instrument for public funds to be used as part of a $10 million loan through the Japan Development Bank (JDB) for the construction of a new plant facility in Japan. Such financing, if approved, would serve several objectives: (1) it would answer foreign outcries of unfair subsidies and discrimination; (2) it would encourage direct foreign investments; and (3) it would promote industrial development in outlying regions. The JDB, as of June 1982, had made loans of more than $25 million (total) to five joint-venture companies in which foreign interests held a majority of shares, and it had also extended loans to joint ventures in which foreign interests held less than 50 percent shares. Realizing that high technology may turn into an arena of serious trade friction, MITI has been trying to solve some of the problems before they get out of hand.

In the United States, some industry and government leaders have talked about the possibility of slowing down the flood of mass memory chips from Japan. The Department of Defense has come under some pressure to invoke an obscure clause (Section 232) of the 1962 Trade Act, which would authorize the suspension of foreign imports if national security is jeopardized. Since the semiconductor industry is essential to America's national security, some industry spokesmen feel that protecting domestic producers from a severe "shake-out" caused by a deluge of Japanese imports would be justified in terms of the country's legitimate concerns. National security also happened to be the reason cited in AT&T's decision to grant Western Electric a fiber optics contract despite Fujitsu's lower bid.[17]

Japanese executives interviewed, however, expressed confidence that the U.S. would not resort to protectionism, because the U.S. computer industry (among others) would be hurt as much as Japan. A large number of U.S. computer firms, they reason, depend upon cheap, reliable semiconductor components from Japan for installation in their terminals. If the influx of Japanese parts were slowed down, there would be either a shortfall in supply or a rise in component costs, both of which would be damaging to their competitiveness. Japanese executives assume therefore that these computer companies would resist protectionist pressures. Such reasoning makes sense in theory but underestimates the complexity of the political processes in Washington.

Interestingly, some American leaders make equally naive assumptions about Japan. They believe that the application of strong outside pressure is the only way of effecting changes in Japanese attitudes and policies. Japan has no option but to comply, they feel, because Japanese companies are more dependent on the U.S. market than the other way around. Restricting their access to the U.S. market would hit them where they are most vulnerable; hence the threat of retaliation, they think, can be used to force Japan to abandon unfair trade practices and open up their market.

Whether the issue is reciprocity or equal market access, the U.S. semiconductor industry insists that its commitment to free trade is unwavering. In view of its current world dominance, this stance undoubtedly makes sense; free trade in a multilateral framework is in America's national interests. But if national security is brought into the picture, as hinted in recent discussions, the possibility of regulating the influx of Japanese imports at some point in the future cannot be discounted. Because of America's pivotal role in maintaining the world's military balance, government leaders may not be willing to let its traditional supremacy in high technology industries be seriously eroded.

The semiconductor industry has also used the Japanese threat to persuade leaders in Washington of the urgent need for policy measures that provide American firms proper incentives to maintain their growth and competitive staying power. It is widely believed that the Japanese government bestows certain advantages on Japanese firms that are hard to duplicate in the U.S., such as subsidized support for national research projects, low-interest loans for high-priority industries, and "targeted" industrial policies. To neutralize such state-bestowed advantages, the U.S. semiconductor industry has asked for, and received, supply-side relief in terms of accelerated depreciation allowances, R&D tax exemptions, lower capital gains, and other provisions. It has also gained initial Justice Department approval for research cooperation involving twelve member firms of Microelectronics and Computer Technology Corporation (MCC). In response to MITI's organization of the Fifth-Generation Project, DOD has organized its own supercomputer project, an ambitious, multiyear undertaking designed to do everything the Japanese have set out to do, at an estimated cost of upwards of $1 billion. Government and industry are also pumping more money into university-based research in hopes of pushing back the frontiers of basic knowledge and of turning out highly trained graduate students and young faculty.

Such developments, undertaken in response to the Japanese challenge, will help to offset some of the advantages of Japanese industrial policy. It adopts some of the very same measures: supply-side incentives, cooperative R&D, exemptions from anti-trust, R&D subsidies, and even government research projects in "targeted" technologies. In some areas, in fact, state support even goes beyond that associated with "Japan, Inc.": for example, industry-university research and cooperation, large defense-related procurements, patent restrictions on R&D done under government contract, classified research, and so on. Some Japanese have described it all cynically as "America, Inc."

Though there have been adjustments in America's R&D effort, the U.S. semiconductor industry is not united on the question whether it needs the full array of industrial policies to compete effectively with Japan. It is apparently not interested in forging the kind of close, symbiotic ties with government that it believes exist in Japan. Nevertheless it would be pleased if government-business interactions ceased being needlessly adversarial and if an environment more hospitable to the growth of high value added sectors could be created.

The Pursuer Industry: Coming from Behind in Japan

We have seen that the government played a central role in reducing the high costs and risks associated with starting the semiconductor industry in the United States. In Japan, which mobilized in earnest after the U.S. had already established a seemingly insurmountable lead, circumstances dictated an even greater role for the government. A. Gerschenkron and others have pointed out that the urgency of closing the gap as quickly as possible so as to safeguard military, political, and economic sovereignty usually forces latecomer states to assume heavier responsibilities than firstcomer states.[18] Japan's semiconductor industry is more accurately characterized as a "pursuer" industry rather than a "latecomer" in the chronological sense of having started late: though several Japanese electronics firms were doing semiconductor research long before the formation of Intel, Advanced Micro Devices, and other U.S. companies, the Japanese semiconductor industry as a whole quickly fell behind in technology and world market share.

Catching up required strategy. Since Japan has only a modest military establishment, operating on a miniscule procurement and R&D budget, the American pattern of development was obviously out of the question. The Japanese government had to devise a strictly civilian strategy, using a nonmilitary set of policy instruments. To be sure, those instruments of state were designed to serve basically the same functions as military backstopping; namely technology push, demand pull, and cost and risk reduction. But Japan's overall strategy and tactics had to be different, given its come-from-behind, pursuer status, its lack of a large military establishment, and differences in the institutional context of government-business relations.

The "latecomer" literature is replete with examples of how follower countries like Japan have managed to exploit the advantages of late entry. There is no doubt that it is in some ways easier to start second: start-up costs tend to be lower, there is less aimless groping; the costs of basic research and development are a fraction of those incurred by firstcomers; technological "leapfrogging" is possible; and aggregate growth rates tend to be faster (unadjusted logarithmically). On the other hand, for newly emergent industries like semiconductors, which are technologically volatile, the penalties for late entry can be substantial. Despite heavier initial outlays, firstcomers have first crack at emerging markets and the opportunity to garner a dominant share of the market virtually unchallenged.[19] Moreover, pioneers can accumulate considerable production experience and know-how, a clear advantage in terms of moving rapidly down the learning curve.[20] Thus, pursuers can find themselves caught in a hopeless bind: late in getting started, far behind in technology, forced by circumstances to concentrate on products at the lower end of the value-added spectrum, lacking scale economies of production, threatened by foreign imports from firstcomer firms, faced with competition from foreign subsidiaries operating in home markets, and constantly under the threat of being overwhelmed.

Under such circumstances how do pursuers cope? How, for example, has Japan managed to overcome these formidable barriers? The standard repertoire of instruments used by the U.S. government to help create the semiconductor industry—technology push and market demand—are usually not enough to propel pursuers to the forefront of world competition. The Japanese government, in consequence, has helped to speed up the tempo of development by relying upon three key mechanisms: (1) close government-business communications and cooperation through a variety of institutional channels; (2) financial facilitation through budgetary allocations, regulation of financial institutions, and monetary and fiscal policies; (3) integration of semi-conductor industry policies within the broad framework of a comprehensive industrial program.

Japan's capacity to bring all three ingredients to bear in the development of its semiconductor industry sets it apart from other aspiring but unsuccessful latecomers; it also distinguishes Japan's approach from America's. The key to Japan's strategy has been its emphasis on incentives aimed at heightening, not stifling, commercial competitiveness.

Of the three mechanisms the second—financial facilitation—may have been, in the early phase, the most significant. The first—government-business cooperation—is of comparable importance, since public—private sector consensus can accelerate the pace of industrial catch-up. The third mechanism—a comprehensive industrial program—grows out of, and is closely tied to, the first two; it is adopted in order to ensure that money, manpower, and resources are channeled toward the high-priority, high-growth areas.* Not all pursuers are capable of mobilizing the three mechanisms. Thus Japan's ability to draw upon all three helps to explain its success in playing the catch-up game. One of the distinguishing characteristics of Japan's "pursuer" system is that state power can be applied flexibly over a period of time in response to changing circumstances and the industry's life cycle, which usually means active assistance and support during an industry's infancy and early growth, phased withdrawal during maturity, and careful monitoring, coordination, and regulation during the period of decline.[21] Even after the difficult first phase of high market and product uncertainty is past and the industry has not only caught up with the front-runners but is headed toward full maturity, two of the three mechanisms—government-business cooperation and a comprehensive industrial policy—can continue to be used to great advantage. If America's military-industrial system has been well suited for the creation of high technology industries, Japan's "pursuer" system has been well designed to facilitate the task of industrial catch-up.

Japan's catch-up system is in some ways a "mobilization state" (though not in exactly the same sense in which D. Apter has used the term).[22] Postwar Japan has been geared to achieve urgent national ob-

---

* MITI officials maintain that the notion of "targeted industries" so frequently criticized in America is outdated and inaccurate. Their explanation is that industrial sectors, not to mention individual firms, are not "targeted" for accelerated development; rather, seminal technologies and high-growth areas, such as semiconductors and biotechnology, are identified as high-priority items from the standpoint of Japan's economic growth and well-being. These priority technologies cut across industrial sectors.

jectives and organized to channel resources toward those goals, with
the central government in command of an assortment of effective administrative instruments. Looking back on Japanese history, one can
argue, in fact, that Japan has been in a perpetual state of mobilization
since its opening to the West in the mid-nineteenth century. Despite
many major differences, there are rather striking threads of continuity
linking the postwar to the prewar and wartime periods. E. Sakakibara
and Y. Noguchi point out that Japan's postwar government carried
over some of the same institutional instruments of financial control
from the period of wartime mobilization.[23] With comparatively far-
reaching powers at its disposal—over the setting of interest rates,
credit allocation, foreign exchange, international capital flows, international trade, technology licensing, and so on—Japan's economic
ministries were able to direct the course of Japan's economic recovery
and growth. Shunning military or national security goals, the postwar
Japanese government galvanized the country to pursue economic objectives. Indeed, Japan's national security came to be defined almost
exclusively in terms of purely economic interests. In the metamor-
phosis from military to merchant state, Japan's hold over some of the
key mechanisms of centralized coordination and control has enormously enhanced the state's capacity to steer Japan on its course of
industrial development.

The government's extraordinary control over the economy was possible because Japan had been insulated from the international economic system in such areas as trade and capital flows. As Japan's economy has "liberalized" under mounting foreign pressures, the intricate
package of central controls has begun to unravel. MITI has already lost
most of its power to regulate technology transfers, deflect foreign direct investments, and license raw materials imports. The Ministry of
Finance (MOF) is slowly but surely losing its power to set interest rates
and restrict capital movements. Obviously, these changes have diminished the government's capacity to guide Japan's economy; but the
state still retains a significant degree of power because of industry's
need to adjust continually to changes in the international environment. International mediation, especially with the United States, is indeed one of the government's central functions from the standpoint of
Japan's semiconductor industry. Trade conflicts and world financial
problems have helped to prevent government power from ebbing
faster than internationalization might have dictated.

The presumption in America is that the government should not tamper with the market mechanism unless circumstances absolutely force

it to. This disposition gives government behavior a lurching or schizo-phrenic cast when the state is forced to respond to interest group pressures to intervene. In Japan, the state is expected as a matter of course to do whatever is appropriate and necessary to promote industrial growth and prosperity.

Although the Japanese government adheres wholeheartedly to the basic tenets of capitalism, it does not view its role in "either-or" terms: that is, either "hands off" or reluctant intervention. The boundaries between public and private sectors are not as rigidly dichotomized. As the state is the guardian of national interests, government intervention does not carry the same stigma often associated in America, for instance that it is the prime cause of inefficiency, an act of last resort when the market malfunctions.[24] The Japanese state's fundamental commitment to facilitate industrial development underlies the broad range of its policy actions and reduces the schizophrenic and political cast of government behavior.

Government promotion of the semiconductor industry. To come from behind and catch up in semiconductors, the Japanese government has had to rely on a varied menu of policies, some of which had been used to promote growth in "smokestack" sectors, and others of which were intended to serve the specific needs of the semiconductor industry. The list of policies has included:

1. Central coordination and consensus-building:
    —General agreement on industry-wide goals involving MITI, industry, and LDP
    —Formulation of a "vision" for semiconductor development and its integration within a comprehensive industrial policy
    —Coordination of, and subsidies for, large-scale national research projects of high priority for the industry as a whole
    —A rough division of labor in research effort between government laboratories (basic research with high costs and risks) and industrial laboratories (commercially oriented research)
    —Extensive information flow and communication exchange between government and industry; elaborate channels for ongoing dialogue

2. Supply-side incentives:
    —Special tax incentives
        Accelerated depreciation
        Research and development tax credits

Deferrals

Special reserve funds to protect against losses

—Japan Development Bank (JDB) and other public and semi-public financial institution loans

—Determination of interest rates; general monetary and fiscal policies

3. Demand-side guarantees:

—Nippon Telegraph and Telephone Co. (NTT) and other public agency procurements

—Indirect measures to boost aggregate growth rates and hence semiconductor demand

4. Infant industry protection (until mid-1970's):

—Import duties

—Control over foreign direct investments

—Control over technology purchase and licensing

5. Legislation:

—Provisional laws for promotion of machinery and information industries, the electronics industry, and information-processing industry

—Selective exemptions from antitrust laws

Several measures, such as tax incentives, public procurements, and subsidies for research and development, are functionally similar to those adopted by the U.S. government; others have no precise counterpart. Of these, special importance can be ascribed to the government's key role in consensus-building and coordination, which has set Japan's pattern of "pursuer" development clearly apart from America's.

Technological catch-up. The first task for a pursuer like Japan in a technology-intensive industry is to lay a solid foundation in research and development on which it can build. The Japanese government has taken an active hand in setting the national research agenda, coordinating the research effort, and supplying funds to reduce the costs and risks of very expensive and uncertain projects. To play this role effectively requires a heavy and constant flow of information between government and industry. The Ministry of International Trade and Industry (MITI), in charge of industrial policy, must serve as a clearinghouse of information. This places it in a position to build consensus, identify research priorities, and oversee research programs.

Vast amounts of information, much of it quite sensitive, are exchanged between MITI and industry through formal and informal channels. Although business leaders are normally very guarded about company information, they are willing to divulge sensitive, sometimes proprietary information to MITI officials in full confidence that it will not be leaked or used against them. Obviously a strong sense of trust binds business leaders to MITI officials. To maintain that trust, MITI officials must scrupulously observe neutrality; they must be perceived to be working for the general interests of industry.[25]

Through painstaking information gathering and exchange, MITI is able to formulate microlevel plans for each industry. This places MITI in a position to link industrial priorities and needs with commercial and technological opportunities, which is precisely the kind of coupling necessary for a technology push to be productive. We have already pointed out that because the United States government largely lacks this capacity, its military-oriented technology push is not always efficiently converted into commercial output and consumer product application.

Out of the countless hours of discussion emerges a "vision" for the semiconductor industry, including a fairly concrete agenda of national research priorities. Institutions like the Industrial Structure Deliberation Advisory Council, composed of representatives from industry, finance, labor, academics, and the mass media, meet to discuss and formally approve the consensus that has been worked out between MITI and industry. The national research agenda is based on an assessment of technological needs and prospects, commercial opportunities, foreign competition, long-term trends, and desired directions.

In 1982 the areas identified as high priority for Japan's technological development included (1) VLSI (Very Large Scale Integrated circuits) research and applications, (2) laser beam technology, (3) software capability, (4) high-speed computational systems, and (5) a fifth-generation computer. Specifically, under the Temporary Law for the Promotion of Specific Machinery and Information Industries (1978-85), three categories of technology have been given priority:

1. Prototype research and development (35 items): integrated circuits, high-performance digital computers, laser equipment, etc.

2. Commercial application and production (12 items): bubble memories, mass storage systems, high performance remote processors, etc.

3. Manufacturing processes (41 items): electronic switching systems, digital computers, etc.

Most of the research and development is carried on by corporate laboratories; but the government is also actively involved through work conducted at its own research facilities, including the Electrotechnical Laboratory under MITI's Industrial Science and Technology Agency (ISTA), and NTT's Musashino Electrical Communication Laboratories; these laboratories have played an important role in basic prototype research such as VLSI technology.

On the basis of extensive consultations, the government also determines which research projects to administer and help finance. Government-sponsored projects tend to be large-scale research efforts, often calling for joint government-industry collaboration. Over the years, Japan's government has organized, coordinated, and subsidized a number of projects, as shown in Table 7.

According to an estimate made by the U.S. Embassy in Tokyo, Japanese government subsidies for advanced R&D in the fields of ICs, software, and computers over the seven-year period, 1976-82, amounted to $355 million, or about $50 million per year; this represented less than 10 percent of total R&D expenditures in these fields.[26] National projects currently under way or on the drawing boards for the near future include:

1. High Speed Computer System for Scientific and Technological Uses, 1981-89. Creation of computers capable of operating at speeds of 10 Gigaflops (a billion floating point operations per second). Budget, $105 million.

2. Optimal Measurement and Control System, 1979-86. Development of a complete system for high-speed remote monitoring and control of large-scale industrial processes, using optical elements for sensing devices and transmission of data. Budget, $81 million.

3. Next Generation Industries Basic Technologies Research and Development Program, 1981-90. Cultivation of revolutionary basic technologies in new materials, biotechnology, and semiconductor function elements. Budget, $472 million; semiconductor portion, $114 million.

4. Software Production Technology Development Program, 1976-82. Development of automated software program modules. Budget, $30 million.

5. Promotion and Development of Technology for Fourth-Generation Computers. Creation of computer comparable to the IBM future series. Phase I: Hardware, 1976-79. Budget, $132 million (government portion). Phase II: Software, Peripherals, Terminal Equipment, 1979-83. Budget, $102 million (government portion).

TABLE 7

Japanese Government Subsidies for National Research Projects, 1966-1983

Period	Project	Approx. subsidy (*million $*)
1966-71	High Performance Computer R&D	42
1972-76	3.75 Series Computer Development	228
1971-80	Pattern Information Processing System	115
1976-80	VLSI Development	150
1976-80	Software Development	28
1979-83	Software for VLSI Hardware	100
	TOTAL	663

6. Research and Development Relating to Basic Technology for Fifth-Generation Computers, 1982-91. Design information processing systems to deal with basic social problems projected for the 1990's. Budget, no precise estimates yet.

Total allocations: $992 million (not including Fifth-Generation Computer Project).

Each project has been designed to propel Japanese technology rapidly forward, in most cases (like software) to narrow the gap with the U.S., and in others (like the fifth-generation computer) to achieve state-of-the-art knowledge. The Japanese government prefers not to underwrite the full costs of national projects, since absorbing all costs and risks can be stultifying and inefficient. But there are national projects, such as the first three listed above, that involve exploratory research of such uncertainty that the government feels it must bear all the costs and risks. Before providing subsidies, the government usually has to have three conditions satisfied: (1) the project must be essential for Japan's technological development and future economic well-being; (2) in the absence of government funds, private firms would probably not undertake the research project; (3) the project must be feasible within the time frame projected and the money available.[27]

Public subsidies, therefore, can be thought of as supplying the "missing element" needed to initiate projects of high capital costs and risks, relatively long gestation, high technological yields, and broad commercial applicability. The government sees it as an indispensable way of advancing the nation's collective good.

Government subsidies take one of three forms: (1) outright grants, (2) low-interest (or interest-free) loans on which repayment is contingent on the success of the project, and (3) low-interest loans that must be repaid regardless of the research results. The government has to be careful about the amount of money given and how it is spent, because it is held accountable for the dispersion of taxpayers' money by the ever present mass media and opposition parties.

The costs and consequences of "targeted" research. There is little doubt that as part of its forward-looking industrial policy, the state's role in R&D has accelerated the process of technological catch-up. This role may have been justified when Japan lagged badly behind the U.S., but the practice of research "targeting" is coming under growing criticism abroad now that Japan has caught up in many areas and is seeking to pull ahead. Does the practice give Japan an unfair advantage? Is it "illiberal" in the sense that it violates the norms on which our international economic institutions and regimes are built? Does it force other countries to follow suit, or risk falling behind? Is Japan being singled out for criticism while European countries engage in similar practices, simply because Japan is so successful? Such questions, now being debated, underscore a point made earlier, namely, that key features of Japan's pursuer system can continue to be used to the country's apparent advantage, even after the catch-up phase is over. When that happens, differences in national systems can turn into sources of conflict, especially if features of one system (however embedded in cultural roots) seem to give one country an unfair edge over others.

In addition to these criticisms, the system of "targeted" industrial policy is also vulnerable on pragmatic grounds. If practiced by more than one country (that targets similar areas), it can lead to worldwide overinvestment and a distortion of capital and labor markets. If only one country implements a targeted policy, its priority sectors could well be strengthened in ways that place the more market-oriented systems at a serious disadvantage. Moreover, it is the large, diversified corporations that benefit most from participation in national research projects. When there is overinvestment and high levels of market concentration, as a consequence of targeting, such structural circumstances can lead to the kind of cartelization long evident in Japan's steel industry.

Another criticism leveled at targeting technologies has to do with trade repercussions. The effect of participation in national projects, critics argue, is to reduce disparities in technological capabilities between Japanese firms as well as to collapse the time spread in bringing products to market. By the time a project has ended, companies are fairly evenly positioned to start the race for commercial applications. A number of large firms tend therefore to come out with similar products at about the same time, leading to "excessive" competition at home and an aggressive search for markets abroad. This results in a phenomenon Japan's trade partners have described as an "export torrent" or "export downpour" (*shuchu gou*). If sustained over a period of time, such "torrents" can severely erode the position of foreign firms in their own home markets. For countries already hard hit by unemployment and corporate bankruptcies, like the United States in 1982, "export deluges" can aggravate painful problems of adjustment. For all these reasons, therefore, many leaders in the U.S. semiconductor industry have objected strenuously to Japan's continuing reliance on targeted industrial policies; many feel it is a carry-over from an earlier era of infant industry protection and is no longer justifiable on grounds of technological backwardness.

The standard Japanese rejoinder is to assert that technological targeting violates none of the extant GATT rules of trade and is therefore perfectly consonant with prevailing norms of fair competition. Other countries in Europe have followed similar policies for years, with the hand of government much heavier than in Japan. The only reason Japan is singled out for blistering attack is that the system seems to work in Japan and its successes threaten the technological and industrial position of other countries. If the United States feels that technological prioritization gives Japan's semiconductor industry an "unfair advantage," there is nothing to prevent it from adopting similar practices.

The Japanese do not deny that "torrential exports" have inundated foreign markets in certain product areas, but the way they explain it, the cause is not technological targeting but rather the intensity of competition in Japan. Since 1977-78, MITI has undergone a major transformation—from encouraging exports ("export or die") to facilitating imports in order to soften foreign criticisms—as a result of Japan's embarrassingly large trade surpluses. It apparently identifies a more balanced bilateral current and trade account as lying in Japan's national interests, since trade frictions, U.S. pressures, and the threat of retaliation might intensify otherwise. In the case of 64K RAMs, for example, MITI officials constantly reminded Japanese semiconductor manufacturers of the volatility of the trade situation in the United States; they emphasized that, because the 64K RAM was being used as a kind of barometer of Japan's trade behavior, caution and restraint had to be exercised. An "export downpour," they feared, would trigger a protectionist backlash. Such warnings did not prevent Japanese firms from winning over 70 percent of the U.S. merchant market in 64K RAMs (1982), but it may have restrained what otherwise might have been an ever greater deluge. When trade frictions did develop, one MITI official acknowledged that "MITI may have to become a tougher traffic cop in slowing down speeding exports, even though export traffic is very hard to control." [28] Not a few Japenese businessmen, chafing under MITI's traffic signals, have come to feel that, far from unleashing an export deluge, MITI since 1978 has been reining in Japanese export companies and trying to facilitate and expand foreign imports. [29]

To the charge that targeting distorts the market mechanism, the Japanese reply that, quite the contrary, it actually enhances market competition and ensures that no single firm will monopolize the fruits of innovation. By pushing back state-of-the-art frontiers, national R&D projects contribute to the world's pool of knowledge and hasten the pace of technological development. This accelerates economic output and raises the world's level of material well-being. Therefore, even though foreign competitors object, the Japanese maintain that the world is better off whenever technology is advanced, whether it is achieved through targeting or not.

Research and development trends in government support. During the fifties, when the U.S. government and the Bell Laboratories were investing in semiconductor research, neither the Japanese government nor Japanese private enterprise was prepared to assume the high risks associated with R&D investment. The government had no pressing national security reason to do so, and industry did not have the means to invest at a rate that would have cast Japan in the role of pioneer. [30] This failure reflected shortsightedness, to be sure, but we must remember that Japan at that time was very short on capital to meet the multiple needs of its war-weakened industrial economy. It was not until the early sixties that the Japanese government, after seeing what the United States had developed, came to appreciate fully the significance of semiconductor R&D; then years passed before that recognition was converted into the allocation of significant sums of money. By the late sixties and early seventies, the United States had already opened up a huge lead, forcing Japan to resort to the latecomer's strategy of buying patent rights to narrow the gap.

Even with the burden of royalty payments, Japanese firms were able to make dramatic advances without paying the long-term price of seeing foreign competitors establish permanent footholds in their own market. Obviously, if they had not had access to patent rights, the pace of technological progress would have been much slower, more uncertain, and perhaps more expensive. Japan's semiconductor industry has come a long way in a surprisingly short period of time by following a strategy of technological "leapfrog."

Now that the Japanese have come to a stage where they can no longer push ahead simply on the basis of buying technology but must demonstrate that they can innovate, the government has decided to set aside increasing sums of money for research and development. Total R&D in Japan currently stands around 2.2 percent of GNP, up from 1.2 percent in the early sixties. Of the total, the government expenditures account for only 29 percent, compared with over 50 percent for the United States, the United Kingdom, and France, which have much larger defense budgets.[31]

Because innovation is indispensable to Japan's industrial future, however, the government's share of a significantly larger R&D effort is going to be increased. It is expected that total R&D expenditures will rise from 2.2 to around 3 percent by 1990 with government spending expanding from 27 to around 40 percent. In absolute amounts, the increase will be far greater than the percentages indicate, if Japan's economy grows at a rate of around 3 or 4 percent annually. At the very time when massive infusions of capital for nonimitative R&D are needed, therefore, Japan's "catch-up" system will be pumping more and more into the technological pipeline.[32]

In 1982, as pointed out earlier, MITI launched the Research and Development Project of Basic Technology for Future Industries, in order to advance basic R&D in high technology industries, including microelectronics.[33] Under development are super-lattice devices for high-speed calculation under normal temperatures, three-dimensional integrated circuits for multifunctional and large-capacity performance, and fortified integrated circuits for reliable performance under adverse circumstances.* Like the VLSI project, this ten-year program signals a shift in government orientation from applied research to more basic research. MITI will provide material support and various incentives, but most of the actual research will be carried out by private industry, following a "parallel development" formula in which a number of different R&D approaches are taken to solve a common task or problem. The research will be continuously evaluated over the ten-year period,

---

* If successfully completed, these devices will upgrade the performance of Japan's commercial and military electronics equipment.

and if successfully completed, the project will give Japan's semiconductor industry added momentum in its drive to catch up with and overtake America's leaders. It would appear therefore that the Japanese state is doing a better job for its semiconductor industry than its U.S. counterpart.

Demand side boosts. We have noted the importance of "demand pull" in laying the early commercial foundations for America's semiconductor industry. Guaranteed sales of integrated circuits for Minuteman missiles and the Apollo spacecraft, in fact, gave the fledgling semiconductor industry a greater boost, arguably, than government efforts at "technology push," because it provided a stable production and financial base on which to develop economies of scale and to move down the learning curve. If guaranteed demand has served so valuable a function in the U.S., what, if anything, has the Japanese government done in the way of "demand pull" measures? Indeed, what options have been open, given the fact that Japan, a merchant state, has not had Minuteman or Apollo programs by which to generate demand?

Finding a functional equivalent for military demand pull seems at least as necessary for pursuers as it is for pioneers. Countries that find themselves in "come-from-behind" situations, as Japan did, cannot afford to let pacesetters take full advantage of their lead by allowing their own market to be dominated. Somehow, "come-from-behind" countries must find ways of building scale economies of production so as to get down the learning curve as fast as possible. It is no easy task. One of the stumbling blocks to the development of a world competitive semiconductor industry in Europe has been the inability of European governments to provide powerful "demand pull" stimulus through public procurements during the early phases of the industry's evolution.

The initial disadvantages of starting late or falling behind can, of course, be overcome by joint ventures with front-runner companies or participation in multinational arrangements. That sort of strategy can bring an immediate infusion of technology, capital, product portfolios, captive demand, and marketing outlets. Under optimal circumstances, countries can take advantage of such transfers to develop their own indigenous infrastructure. But there is no guarantee that an independent industrial structure will ultimately emerge, or that multinationalization will not lead to the very kind of foreign dominance that latecomer countries want to avoid. Still, for some countries, the gap with firstcomers is so wide that they have no other alternative.

The Japanese chose to pursue the path of indigenous development, even though they knew the catch-up process would take longer and would mean sacrificing the built-in benefits of technological and organizational diffusion associated with direct foreign investments. They believed that, over the long run, the advantages of having control over their home market would outweigh the short-term disadvantages, if they could get the semiconductor industry off the ground and in a position to compete. This choice could be made because Japan's bureaucratic administration had the authority effectively to restrict foreign investments and protect infant industries. Such measures constituted, in effect, a form of demand-side assistance; for, by keeping foreign manufacturers out of Japan and raising the price of imports, domestic producers had a much better chance of selling their wares to domestic consumers.

Japan's tariff rates for ICs stood for many years at 15 percent, lower than the European Community's 17 percent but higher than America's 10.1 percent. In 1972, Japan's effective rate dropped to 12 percent; by 1980, as Japan's industry became more competitive, the rate was brought into line with America's at 10.1 percent. It was reduced further in 1982 to 4.2 percent, several years ahead of schedule; and the expectation is that by 1987 it will be eliminated altogether. By contrast, tariff rates remain high in Europe at 17 percent, with no schedule for reduction. Japan no longer shelters its semiconductor industry from foreign competition by means of tariff barriers. Controls over direct foreign investments have also been lifted, removing another pillar of protectionism. Where formal barriers are concerned, therefore, Japan maintains that it is as open as any country; but though this may be true today, the government has had a record of infant industry protection and kept it in place longer than the competitiveness of their industries probably warranted.

Despite the fact that it has lacked large defense and aerospace industries, the Japanese government has still managed to boost demand for semiconductors through (1) effectual macroeconomic management, which has fostered high-speed growth and the rapid development of the consumer electronics industry; (2) procurements by Nippon Telegraph and Telephone (NTT), a public corporation in control of Japan's booming telephone and telecommunication needs; and (3) the creation of Japan Electronic Computer Corporation (JECC), a computer leasing company jointly owned by industry and government, which filled the role of purchasing computer equipment from manufacturers and leasing them to end-users. Taken together, this package generated enough demand to accelerate the development of the fledgling semi-conductor industry.

Assured sales to NTT helped Japanese supplier firms in ways similar to that of defense and space contracts in the United States, that is, by support for engineering design and development, cost reduction for production and testing equipment, immediate and guaranteed recouping of partial R&D investments, scale economies of production, movement down the learning curve, and the meeting of exacting standards of performance and reliability. Data on comparative price levels are unavailable, but it would not be surprising if NTT paid higher prices than commercial purchasers. It should be recalled that the average price of transistors for U.S. military procurement was significantly higher than commercial rates (Fig. 4), constituting in effect a kind of "cost cushion" or indirect subsidy for research and development. Even without military and space programs, therefore, the Japanese government found ways of nurturing the semiconductor industry through the demand pull of public procurements.

So far as timely demand-side facilitation is concerned, JECC may have played nearly as significant a role as NTT.[34] The creation of JECC in 1961 gave Japan's seven leading computer manufacturers the financial wherewithal to establish a solid foothold in the growing computer leasing market. Forming a semigovernmental joint venture made it possible for the manufacturers' debt capacity to be greatly enlarged beyond that of any single company. The Japan Development Bank could become its largest single creditor, supplying funds at lower than commercial interest rates. JECC began with capital of $3 million and grew annually at a rate of over 65 percent; by 1970, scarcely nine years after its inception, JECC had purchased $250 million worth of computer equipment. By 1981 the figure had climbed to over $2.2 billion with 2,332 varieties of systems; the statistics are even more impressive in terms of cumulative amounts. By March 1981, JECC's cumulative purchases had exceeded $7.25 billion with rental revenues of $5.6 billion.[35] Though it has had problems, some of them by-products of its successes, JECC has been very instrumental in helping Japan's information-processing industry carve out a secure market position in the face of stern foreign competition, particularly from IBM. Without JECC, Japanese manufacturers would have had a much harder time.

The importance of the information-processing industry for semi-conductor production in Japan cannot be overstated.[36] Except for Oki and Matsushita, the largest semiconductor makers are also computer manufacturers. The Japanese government views the computer industry as critical for its knowledge-intensive economy of the future. Indeed, semiconductor production is deemed of highest priority, precisely because it is so central to the competitiveness of Japan's computer industry. To the extent that JECC has helped Japan's computer manufacturers defend its home markets against IBM, it has promoted semiconductor production and sales over the short run and laid the foundations for the country's whole information-processing industry, one of the biggest and most important end-users of semiconductor components in the long-run future.

In short, with respect to those areas where Japan's semiconductor industry needed help most—demand pull, coordination of technology push, protection against firstcomer dominance, scarce resource mobilization, financing and promotion of key end-user industries like computers—the Japanese government has come through with indispensable assistance.

Antitrust. Compared with U.S. standards, the Japanese government's stance on antitrust seems to rest on a looser interpretation of the letter of the law and a more pragmatic application. Firm size, product prices, and market concentration are not the sole criteria on which antitrust decisions are based. Such considerations, though obviously central, are weighed against other criteria such as international competitiveness, production efficiency, world market share, industrial orderliness, technological virtuosity, and adaptation to shifting comparative advantage. Large market share, by itself, is not necessarily *prima facie* cause for divestiture proceedings.

One of the striking differences in antitrust attitudes between Japan and the United States is the strong faith placed by Americans in the principle of free market competition as the essential mechanism for advancing the public good and preserving the democratic processes, whereas Japan, which has found itself in a catch-up mode for over a century, has placed greater stress on economic output and pragmatic results. This is not to imply that Japan's Fair Trade Commission (FTC) is unconcerned about price-fixing or corporate collusion in restraint of trade; clearly its mission is to protect the public from such practices. But the overall setting within which antitrust is administered, including Japan's institutional patterns, has given it a pronounced economic and national interest cast. In other words, there is in Japan an inherent tension between antimonopoly laws, the legacy of America's occupation, and certain deeply ingrained administrative practices. Viewed from a broader perspective, the problem goes even further: it is nothing less than the government's role in microindustrial management. How far should the visible hand of state be allowed to extend? What is the proper place of antimonopoly legislation in a pursuer system oriented toward the achievement of high-priority economic objectives?

Unlike the situation in the United States, prominent features of the Japanese government's role are themselves an integral part of the problem. Antitrust is not simply a matter of reining in the abuses of the private sector. MITI's capacity to forge consensus within industrial sectors and to coordinate policies between industry and government may lead to potentially undesirable consequences—like setting production levels or prices. When such abuses occur in the United States, they emerge out of the structure of market concentration or factors embedded in the private sector. Hence, what is a definite asset for industrial harmony and orderliness in Japan—namely, close government-business cooperation and the power of the economic bureaucracies—turns out to pose serious potential conflicts where antitrust is concerned.

Consider administrative guidance as a case in point. From a government-industry point of view, administrative guidance—an informal non−legally binding set of guidelines—is a convenient and effective tool of microindustrial management. For MITI, it provides a flexible way of securing voluntary compliance from industry without having to go through the legal channels of the Diet. Averting political entrapment is important for MITI, because of its desire to preserve maximal autonomy and flexibility of operation. If MITI had to funnel all aspects of industrial policy through the Diet, it would quickly find itself dependent on party politicians, vulnerable to partisan politics, and engulfed in pressures from vested interest groups. By relying on administrative guidance, MITI can deftly sidestep the kind of excessive politicization that undermines bureaucratic control and opens the doors to partisan lobbying. From the perspective of private industry, guidance provides a mechanism for building consensus and using government authority to sanction it; here again, the overlayer of legalistic red tape is neatly avoided and government-business relations preserve an enviable degree of pragmatism and flexibility.

Antitrust complications arise when administrative guidance is used to moderate "excessive competition" or regulate investment levels, as is sometimes the case for structurally or cyclically depressed industries. The reason MITI "suggests" investment levels for certain industries—when it adheres so staunchly to market principles in most other circumstances—is that it wants to maintain structural orderliness, industrial harmony, and stable prices. Although antirecession cartels help certain industries avert fratricidal competition, they raise profound legal questions, particularly about possible violations of Article 9 of the antimonopoly law.[37] In a landmark 1974 decision, in fact, the Tokyo High Court ruled price fixing in the petroleum industry illegal and stopped short of declaring administrative guidance illegal but noted that limits had to be imposed on its use.

More stringent antitrust enforcement makes it hard to imagine projects like the VLSI being organized in America. Although permission is granted every year to a small number of U.S. firms undertaking joint research, none of these projects even remotely approaches the VLSI in scale, ambitiousness, and close level of cooperation between industry and government. The VHSIC project is qualitatively different in that it has not established cooperative laboratories under a semigovernment research association where research scientists from rival firms and government laboratories come together to work. In the United States, the normal practice is for the federal government to award research contracts to individual firms through a process of competitive bidding.

The different approaches to antitrust taken by the U.S. and Japanese governments have different consequences for the two semiconductor industries. In the past, America's antitrust policies have kept corporate giants like Western Electric from selling on the merchant market while at the same time imposing compulsory licensing on Bell Laboratories. These policies have had the effect of stimulating technological innovation. Barriers to new entry have been kept low and the emergence of many small, new companies oriented toward state-of-the-art R&D (in conjunction with the Bell Laboratories' ongoing work) have created an atmosphere and industrial structure of enormous dynamism. Court rulings and pending legislation may change the nature of that environment; depending upon what happens in the courts, Congress, and the Executive branch, antitrust enforcement could conceivably deprive the U.S. industry of some of the dynamism that cast it in the role of world pioneer.*

Japan's more pragmatic enforcement of antitrust has given the state greater leeway to mobilize and coordinate finite resources. Antitrust provisions exist in a relationship of latent tension with the govern-

---

* A variety of bills were introduced in the U.S. Senate and Congress in 1982 to ensure that joint research is legally permissible so long as competition is not harmed. Senator Paul Tsongas's bill, "R&D Joint Venture Act of 1982," is one example. It is a direct response to joint R&D activities in Japan.

ment's intrusive role in microindustrial management; but neither the courts nor the FTC has done much to shackle such common practices as administrative guidance and national projects like the VLSI. A more lenient interpretation of antitrust seems to serve the needs of a pursuer system. To cast Japan's antitrust policies in broader comparative perspective, however, we should note that in terms of the lenience of interpretation, Japan is closer to the European than to the American model. Indeed, as in other aspects of its political economy, America, not Japan, stands out as the exception.

Summary. In assessing the roles played by the American and Japanese governments in promoting the development of their semiconductor industries, one may be struck, initially, by the variety of differences that seem very fundamental in nature: for example, America's preoccupation with global military security; Japan's obsession with economic and commercial interests; the government's more extensive, ongoing, and industry-specific support in Japan as contrasted to the pattern of early involvement, followed by distancing, and a sporadic but generally laissez-faire attitude in America; cooperative interaction between government and business as the modus operandi in Japan; and various, issue-specific differences, such as antitrust policies and financing. Such contrasts are indeed striking.

On closer examination, however, the underlying similarities become more apparent. Both governments faced similar tasks in the initial phases: namely, to reduce the risks and costs of private enterprise in making a start. In both cases this meant, concretely, that some of the early costs of research and development would have to be borne, and that government would buy some portion of production.

Where the two patterns diverged sharply, as they did over infant industry protection and the organization of national research projects, the divergences could be explained in terms of two circumstances: (1) the two industries were located at different points along the developmental spectrum of world competition, with America the leader and pacesetter and Japan the pursuer; (2) the two political-economic systems were fundamentally different in structures and processes. To compensate for certain weaknesses or deficiencies in market forces, the Japanese government has felt obliged to intervene, whereas the American government could sit back and let the market take its course. Thus, because the Japanese semiconductor industry faced formidable foreign competition, which its American counterpart did not have to contend with until the seventies, the Japanese government felt that it had to protect its "infant" industry through a variety of trade-restrictive measures, including high tariffs and restrictions on direct foreign investments. Similarly, because the country lagged so far behind in technology and had to cope with structural barriers to rapid technological diffusion like permanent employment, the Japanese government has had to take the lead in facilitating intra-industry consensus on research and development and in organizing national research projects, designed to "leapfrog" the state of knowledge ahead. Other differences such as favorable public financing and the use of a conscious industrial policy (discussed in the next section) can also be understood in terms of certain structural deficiencies, such as the early shortage of capital and the underdeveloped state of Japan's equity market, which necessitated some form of government intervention.

Political Structures and Indirect Policies

Discussion has focused, to this point, on what the governments have done specifically to promote their semiconductor industries. Just as important are the general policies and structural factors that have had an *indirect* but equally crucial bearing on industrial competitiveness. This section will examine the indirect factors, particularly those embedded in the structures of the two political systems: specifically, electoral dynamics, politicization and public policy, government institutions, industrial policy, and the political influence of the semiconductor industries. Let us begin by looking at the relevance of elections, since public policies are inevitably conditioned by the imperatives of electoral politics.

Electoral stability. For semiconductor and other industries, the value of coherence and consistency in the political management of the economy cannot be overstated. Conducting business on a day-to-day basis, not to mention planning for the long-term future, can be greatly facilitated by predictability and evenhandedness in the way the government runs the economy. Fortunately, for both American and Japanese semiconductor industries, the electoral politics that affect macroeconomic management rank among the world's most stable. The postwar dominance in Japan of the conservative, pro-American Liberal-Democratic Party (LDP) is particularly noteworthy. Except for one brief interlude of coalition rule, the LDP has commanded a majority in both houses of the Diet for virtually the entire postwar period, a record unprecedented among the large industrialized states.[38]

As a result, the political environment in postwar Japan has been about as favorable to business interests as in any major country in the world. The LDP-led government has pursued a set of policies that have consistently encouraged productive growth, national economic interests, enhancement of material well-being, and producer over labor interests. Government policies in Japan have come closer than in almost any other large industrial state to the old adage, "What's good for business is good for the country." Japanese corporations have fared well in terms of tax incentives, interest rates, financing, control over inflation and unemployment, labor relations, spending priorities, macroeconomic management, and systematic advancement of national economic interests overseas; and the pro-business orientation of the LDP and central bureaucracy has had a great deal to do with that.

In the case of the semiconductor industry, such environmental factors can affect the entire range of corporate operations, from production costs to market demand, competition, and profitability. To the extent that growth in semiconductor demand is related to rises in income (income elasticity), Japan's semiconductor industry has benefited greatly from the rapid expansion of aggregate output. From 1978 to 1980, Japan's economy grew at a significantly faster rate than America's: in 1978, 5.0 percent compared with 4.6 percent; in 1979, 5.5 percent compared with 2.8 percent; and in 1980, 4.2 percent compared with -0.2 percent. Percentage comparisons are not altogether accurate because America's economy is twice as large as Japan's; nonetheless, the growth of demand especially in Japan's consumer electronics industry has pulled the semiconductor industry toward ever greater expansion.

America's federal system, featuring fairly regular alternation of power between the Democratic and the Republican parties, may pose greater difficulties in terms of effectual, nonpoliticized management of the economy. Although periodic "housecleanings" in the executive branch may be desirable from the standpoint of democratic politics, they make continuity and coherence in policy orientation difficult, particularly with respect to economic management. A new president or a change in party control of Congress usually means a shift in economic policies. In America's two-party system, policy "U-turns" are hard to avoid. Fluctuations and reversals in capital gains taxes, to cite just one example, have undermined investment incentives, making it difficult to sustain investments continuously at high levels.

In addition to regular shifts in policy direction, which can complicate long-term corporate planning, the politics of presidential elections can also induce incumbents to manipulate short-term economic trends in ways that attempt to maximize chances for reelection. One study has shown that in a majority of democratic states, including the

United States, a phenomenon called an "electoral economic cycle" is commonplace. Its aim is to expand real disposable income and lower levels of unemployment prior to election campaigns. Tight money policies might be relaxed, or federal payments for welfare and other transfer payments may be stepped up. The regularity of elections means it is hard to sustain a very tight monetary policy; without the political will to control money supply or spending, it is tough to break inflationary expectations and keep price levels stable over time.*

Because of the politics of elections, federal expenditures in highly visible areas like welfare and defense are hard to reduce; it is often easier politically to increase allocations. Entitlements and military outlays tend to be a net drain on industrial productivity, whatever windfall benefits may go to individual semiconductor companies. Huge deficits, incurred over decades, have often led to a "crowding out" of capital for private sector investment, thus sapping the economy of some of its potential dynamism. Japan has not had the burdens of heavy defense and welfare spending for most of the postwar period.[39] In sum, the long dominance of the LDP, and the politics of fiscal and monetary policies have helped to create an environment extraordinarily favorable to business interests.

The LDP's "grand coalition": support group bifurcation. How has the Japanese government managed to sustain a climate so favorable to business interests without succumbing to parochial pressures from interest groups or to partisan politicking? An analysis of the LDP's "grand coalition" of support, and the nature of its interactions with its support groups, tells us a great deal about the political economy within which Japan's semiconductor industry operates.

The LDP draws support, in varying degrees, from nearly all sectors of society, but its interactions with two broad groups deserve special mention: (1) the "labor intensive" sectors of society—farmers, fishermen, small-medium entrepreneurs, retailers, doctors, those in the services, the elderly, and so on—from which the LDP receives votes; and (2) big businesses—the large corporations at the heart of Japan's industrial development—from which it receives financial support. The gerrymandered election system, which allows nonmetropolitan districts to be heavily overrepresented in the Diet, has given rise to a sit-

---

* Edward R. Tufte, *Political Control of the Economy* (Princeton, N.J.: Princeton University Press, 1978). Tufte's time series data indicate that both the U.S. and Japanese governments have succumbed to the electoral-economics temptation. But whether the pattern holds true in Japan or not is open to question. A plausible explanation of the data is that LDP leaders tend to call elections when the economy is doing well, not the other way around.

uation in which powerful interest groups such as agriculture, fishery, and other primary sector occupations have provided the electoral base for the LDP, which, in turn, has converted voter support into pursuit of rapid industrial growth.[40]

From 1952 to 1972, the LDP focused much of its political energy on satisfying its traditional support groups, using the prerogatives of incumbency to funnel subsidies, public works contracts, and favorable public policies to the labor-intensive, largely nonmanufacturing sectors of society in return for votes. The LDP's relationship with these labor-intensive groups has been particularistic, symbiotic, politicized, and unabashedly pork barrel. Not surprisingly, the government ministries in charge of regulating these interest groups, such as the Ministry of Agriculture and Forestry and the Ministry of Construction, have been among the most highly politicized.[*] The LDP used its power to "buy off" these electorally salient sectors by earmarking large sums from the General Accounts Budget (e.g., for rice price subsidies) and by passing generous tax packages (as for medical doctors). In consequence, the "labor-intensive" sectors, particularly those left behind in the processes of industrial development, received the benefits of a continual redistribution of income, and the gap between the "backward sectors" (e.g., the countryside) and the "rapidly advancing sectors" (e.g., the metropolitan areas) never grew large enough to threaten political or social stability. Of course, such stability has come at a price—witness parochial interest group lobbying, political favoritism, corruption, and egregious examples of economic inefficiency in the primary and service sectors. But the damage has tended to be "contained" within single-issue domains,[†] and the most blatant forms of pork barreling and petty politicization have not spilled over into, and contaminated, government-business relations in the manufacturing sectors.

The LDP's relationship with the "capital-intensive" manufacturing sector has been of a sharply different nature: more distant, less particularistic, less politicized, and more firmly mediated by the powerful economic bureaucracies that have jealously guarded their prerogatives against interest-group and LDP encroachments. With the LDP's atten-

---

[*] Other ministries that fall into the politicized category include Transportation, Post and Telecommunication, and Health and Welfare, which regulate narrowly defined fields under which LDP interest groups operate. It is no accident that a number of bureaucrats have been recruited into the LDP as Diet members.

[†] Interestingly, one sees little evidence of interest group coalitions across issues areas; interest groups seem narrowly interested only in issues that directly affect them, hence the term "single issue" lobbying.

tion heavily absorbed in the politics of the nonmanufacturing sectors, MITI and the Ministry of Finance have managed to maintain firm control over policies related to the industrial economy. The LDP has been content to let MITI run the industrial economy, because MITI has demonstrated exceptional competence and the LDP has obtained what it has really wanted, namely, large financial contributions from big business. Japanese industrial policy has therefore enjoyed a degree of insulation from both LDP interference and parochial interest-group lobbying that must be considered extraordinary among postwar industrial states.

Several reasons can be cited to explain this situation. Aside from the peculiarities of Japan's electoral districting, already alluded to, perhaps the most critical factor has been MITI's capacity to forge general consensus with industry. Not only are broad objectives agreed upon within the framework of a long-term industrial vision, but concrete policy measures are also worked out between specific MITI divisions and bureaus (*genkyoku*) and individual industrial sectors.[41] The LDP may try to influence MITI on behalf of special interest groups during the formative stages of policy deliberations, but once an agreement has been arrived at, LDP members have little opportunity to intervene.[42]

MITI and industry are able to arrive at consensus because: (1) a relationship of trust has been built over years of interaction; (2) industry understands that MITI adopts policies in the general interests of industrial sectors and the economy as a whole; (3) MITI's track record over the postwar period inspires confidence; (4) MITI's unusually broad scope of authority compels it to act on behalf of collective interests; (5) industry has had very little to complain about with respect to industrial policies; and (6) MITI has studiously avoided having to rely on legislative action and the law. They firmly believe that the fiercely competitive nature of Japan's company-centered, industrial economy seems to depend on active mediation and coordination on the part of a neutral entity, government, which acts in the national interest.[43] Both MITI and industry prefer that the use of law be kept to a minimum; administrative guidance, as mentioned earlier, is one of the means of sidestepping legislative interference.

The politics of the semiconductor industry. MITI's influence is evidenced by the fact that it has assumed the initiative in organizing LDP Diet members into informal support groups on behalf of various industrial sectors. It has helped to organize two parliamentary caucuses, for example, the Knowledge-Intensive Industries Caucus and the Information Industries Caucus, which together claim a membership of

over 200 Diet members. In exchange for relatively modest political do-
nations, these groups, organized around a core of Tanaka and Fukuda
faction members, see to it that computer and semiconductor interests
are promoted when decisions on such things as tax provisions and special development laws are brought before the Diet. Caucus members
have earned a reputation for being seriously committed to the promotion of Japan's information industries.

Their activities range from gathering information about policies in
the U.S. to building consensus within the LDP and mobilizing support
among members of the opposition parties. When the budget for the
VLSI project was being drawn up, the two caucuses came to MITI's
support in calling for generous allocations from the Ministry of Finance. From MITI's perspective, not to mention the semiconductor industry's, the two caucuses have served a useful purpose not only in
running political interference within the legislative branch but also in
adding weight to MITI's voice within the administrative hierarchy.
Their task is made easier by the fact that few parties stand opposed to
the idea of promoting the information industries. LDP and opposition
parties alike realize that the data processing, telecommunications,
consumer electronics, and semiconductor industries will form the core
of Japan's so-called "knowledge-based economy" of the future.

The situation is very different in the United States. The semiconductor industry has a much longer and harder road to haul in making its
weight felt in the political arena. Not only is the U.S. government less
systematically pro-business, but tailoring public policies to suit the
needs of a particular industry, like semiconductors, is far more difficult
because of the sheer size of the country and the great number of interests, business and otherwise, clamoring for favor. With so many agencies representing diverse interests and points of view involved in the
policymaking process, the SIA must clear formidable hurdles in seeking to shape policy outcomes.

What political assets can the SIA draw upon in its lobbying efforts?
In terms of standard indicators—number of employees, productive
output, geographic spread, years of existence and experience—the SIA
would not seem to command a great deal of leverage. Semiconductor
sales accounted for less than one-half of one percent of the country's
GNP. Employees number only around 180,000, a tiny fraction of
America's 80 million work force. Assembly facilities and corporate
headquarters are located in twenty-five states, but the overwhelming
center of merchant gravity is concentrated in "Silicon Valley," a small
belt running between Palo Alto and San Jose in northern California.

The semiconductor industry cannot count upon a large contingent of senators and congressmen to represent their interests in Washington out of constituency obligations. There are only around 160 merchant producers in the U.S., with the 40 largest representing 95 percent of sales.

But if the crude geopolitical indicators suggest that the semiconductor industry is not among the potentially powerful lobbying groups in Washington, the industry has other things in its favor, most importantly its very nature as a strategic industry. Although its share of aggregate economic output is still miniscule, semiconductor technology is absolutely vital for the whole high technology sector, not to mention for the country's national security. It is also a rapidly growing industry; by 1990, the industry may be producing $40-$50 billion worth of goods. In 1980 it provided key components for the country's $200 billion electronics industry. Semiconductor products, the so-called "crude oil" of the 1990's, have attracted wide attention in the national media. The semiconductor industry is capable of presenting a compelling case for its significance in terms of the country's technological development, economic well-being, commercial competitiveness, and military security. It can appeal, in other words, to a functional, not geographic, coalition of supporters within Congress and the Executive branch.

The SIA is also in a position occasionally to join forces with other interest groups like the American Electronics Association (AEA) if necessary. The two groups worked together in seeking a better package of tax incentives from the Reagan administration in 1981. However, since member companies are not always of one mind, the SIA sometimes finds it difficult to act as a unified body. As it must operate within a climate decidedly less attuned to its needs, the semiconductor industry must bring its case to Washington and before the American public as often as it can.

Industrial policy. One of the structural contrasts most often talked about is Japan's use of a comprehensive industrial policy and America's unwillingness formally to adopt one. Why has Japan rushed so eagerly to embrace an industrial policy when the U.S. government has been skeptical about its effectiveness? What are its major advantages and disadvantages?

To answer these questions requires that we first define the term. Industrial policy refers here to the government's use of its authority and resources to assist or regulate specific sectors or companies. It implies that public policies are consciously differentiated in ways that meet the needs of specific industries or firms. Under this definition, virtually all capitalist states, whether publicly acknowledged or not, implement an industrial policy. Just how extensively it is practiced and what sorts of policies are used obviously vary from one government to another. To highlight the diverse range of policy instruments and multiple set of objectives, a typology of industrial policy has been constructed (see Table 8).

If, by industrial policy, all four categories of the typology are included, one can say that the U.S. government has been reluctant to adopt a comprehensive package. It operates on the conviction that the market mechanism is the most efficient means of resource allocation; the state should not tamper with the market since it cannot pick winners and losers as impartially as the market. To the extent that industrial policy is designed to bend, if not defy, market forces, it will lead to distortions that carry higher costs in the long run than those associated with market-induced adjustments. Industrial policy thus tends to be seen as a substitute for, not a complement to, the market mechanism—and a decidedly inferior one at that.

Despite such reservations, however, the U.S. government has not had the political will to stand back and let the market take its natural course. It has intervened sporadically to protect certain industries, particularly the politically powerful old-line sectors that have gradually lost comparative advantage in the international marketplace. Nor, despite its avowed impartiality, has it treated all industries the same. Some have received benefits not extended to others. Commercially troubled companies like Lockheed and Chrysler have qualified for large, federally guaranteed loans that have rescued them from looming bankruptcy while other firms have been allowed to go under. The government's rate of effective taxation has also fluctuated widely across industrial sectors. Electrical machinery has borne a higher tax burden than have the automobile or shoe industries. And the steel and automobile industries have been given protection from foreign imports while semiconductors have not. All this points to the fact that, notwithstanding rhetorical disclaimers, the U.S. government has in fact backed its way into a tacit industrial policy—one that has protected many of its declining industries from the full force of international competition.

In terms of the industrial policy typology, the U.S. government has taken policies that fall into the first three categories in Table 8—protection, adjustment, and relief; where enhancement of high-growth industries is concerned, it has tended to take a more laissez-faire attitude, believing that these industries are capable of taking care of themselves. Of course, the pull of military demand has had the effect, as pointed out earlier, of giving birth to a number of new high technology industries, and military demand can in a sense be considered the functional equivalent of a forward-looking industrial policy. But if that is so, industrial enhancement has never been the primary objective: on the contrary, civilian spillover benefits have emerged as unintended consequences—almost, in some instances, as accidental by-products. And whether new industries will continue to spin off from the huge military programs is by no means certain. New fields like bioengineering have emerged without any boost from military demand, and, though military applications may eventually be possible, the main momentum behind the industry's development will be provided by the commercial market.

Unlike the U.S. government, Japanese officials have long used industrial policy as a major instrument of industrial growth, resource allocation, and export promotion. Its postwar origins date back to the fundamental decision to develop an industrial structure consisting of the basic raw materials processing industries, heavy machinery and equipment, and machine tools—manufacturing industries that, once developed, would serve as the foundation for precision instruments, consumer electronics, data processing, and other knowledge-intensive, higher value added industries. Although the decision to aim at joining the industrial "big leagues" may have been in some ways risky and overly ambitious, the decision to develop through heavy manufacturing made long-term economic sense, if only because these industries shared important characteristics: high income elasticity of demand, the potential for rapid technological catch-up, and the possibility of steep rises in labor productivity. If carefully planned and diligently pursued, this course of industrial development offered Japan its only hope of sustaining a population of over 100 million at a standard of living that was not only tolerable, but capable of being brought up to the level of the United States and Western Europe. But to realize this bold vision would require long-term planning, identification of early industrial priorities, and the concentration of scarce resources in those areas—in short, a systematic and comprehensive industrial policy.[44]

From its postwar inception, therefore, Japanese industrial policy was seen as a mechanism for overcoming serious deficiencies in Japan's early postwar system, such as an underdeveloped equities market and shortfalls in investment funds. In retrospect, an argument can be made that the government wanted to perpetuate this state of underdevelop-

TABLE 8

Type/Purpose	Policy instruments
PROTECTION	
—Shield domestic firms from foreign imports	—Tariffs; quotas; import surcharges; OMAs; voluntary restraints
—reduce bankruptcies and unemployment	—direct foreign investments in developing nations as means for getting around import substitution barriers abroad
—defy the dictates of shifting comparative advantage for domestic political reasons	
ADJUSTMENT	
—Slow down pace of adjustment to loss of international competitiveness	—Unemployment insurance
	—manpower retraining and reallocation
—acceptance of need to phase out and shift labor and capital into higher value added industries, but only gradually	—regulation of plant investment and capacity utilization
	—antirecession cartels
RELIEF	
—Temporary respite for industries in need of reinvestments and retooling	—Subsidies; bail-outs
	—temporary import restraint
	—government procurements
—prevention of major negative impact on economy	—facilitation of investments and exports
—national security interests at stake	—protection against foreign price competition via trigger mechanism
	—relaxation of regulatory requirements
ENHANCEMENT	
—Encouragement of competitiveness in areas of natural comparative advantage	—Tax relief, R&D incentives
	—capital formation facilitation, guarantees of fair foreign competition
—acceleration of production in high-growth sectors of the economy	—procurements, manpower training
	—guarantees against technological theft
	—foreign exchange rate intervention
	—elimination of foreign trade barriers

NOTE: The above typology is based on the experiences of advanced economies like the United States and Japan. The specific industries affected, and the specific mix of policy instruments, depends on a

A Typology of Industrial Policy

Affected industries	Possible costs
—Sunset industries —labor-intensive low value added, like basic textiles and apparels	—Inflation —labor misallocation —frictions with developing countries —danger of destroying system of free trade
—Industries with world or domestic overcapacity —industries with limited product differentiation like shipbuilding and steel —some consumer durables	—Budgetary drain, retardation of structural adjustments —contagion effect —danger of lapsing into indefinite protection
—Large, basic industries with heavy plant capacity —strong labor unions —large multiplier effects on economy like steel and automobiles —for some countries, agriculture	—Inflationary —market, labor, and capital distortions —expectation of long-term assistance/relief —disincentives for raising productivity and for regaining international competitiveness
—Technology-intensive industries, especially those with high capital costs and short technological and process manufacturing life cycles —high value-added, seminal industries like semiconductors, computers, and bioengineering	—Complaints of favored treatment by other industries or rival firms in other countries —possible frictions with developed areas where one or two countries control large market shares

country's stage of economic maturity; for developing nations, infant industries rather than declining industries may need protection from foreign imports.

ment in order to retain its extraordinary powers,[45] but in the fifties the paucity of investment capital seemed to leave government officials little choice. In order to ensure that scarce funds would find their way to the "right" industries, government officials felt compelled to rely on the allocative mechanism of an industrial policy. Thus Japan's system of control over the banking industry, established after the depression of the thirties and during wartime mobilization, was carried over intact into the postwar period. The government's capacity to regulate money supply, interest rates, deposit rates, dividends, new branch offices, and foreign exchange gave it enormous power, which it used to achieve the goals set forth in the country's industrial plans.[46] Japanese banks became the primary source of capital for corporate financing, and they competed for loans to high-growth industries. And, as Y. Ojimi, a former MITI official, points out, "In judging what makes a growth industry, banks relied on the judgment of MITI as one of their chief criteria."[47]

Japan's industrial policy was consciously aimed at fostering industries with potential for rapid growth. Since it had to rebuild from the ground up, Japan did not have to worry much about protecting its declining or troubled industries. The thrust of its industrial policy was almost the opposite of America's: forward-looking, consciously used, comprehensive, export-oriented, relatively nonpoliticized, and purely commercial.[48] Intervening where it had to but operating within the basic framework of capitalism, the Japanese government saw its ambitious vision for Japan's postwar industrial structure eventually realized.[49]

Just how much of Japan's economic achievements is directly due to MITI's industrial policy is hard to say. Some scholars assign it primary importance; others see it as a causal factor of secondary importance.[50] It is not clear, for example, whether scarce capital would have flowed as fully to the high priority industries, or whether funds would have been diverted to less productive sectors in the absence of an industrial policy. Would Japan's economic growth rates have been significantly slower? And its industrial structure substantially different? On such hypothetical questions, there is no unanimity of opinion.

As seen from the perspective of the semiconductor industry, what are the strengths and weaknesses of American and Japanese industrial policies? Judging from the competitiveness of the two semiconductor industries, the two approaches, as different as they have been, appear to have contributed positively to their development; since causality is hard to quantify, at least one can say that the two approaches have not stunted each industry's development. In the case of the United States, the government's desire to minimize tampering with the market seems to have created a vigorous business climate for the semiconductor industry and the success of the industry suggests that it has been well served by America's decentralized, market-based system of capital and resource allocation. Indeed, it is curious that industry leaders should express a mixture of envy and concern that their Japanese rivals operate within a framework of greater government guidance, particularly in light of the fact that Japan represents the only example of a successful industrial policy in the industrial world.[51]

The American government's reluctance to tailor policies to the special circumstances of individual industries is also based on the fear of opening a political Pandora's box. If policymaking is already highly politicized in America, the chances are that it would become even more so if greater use were made of industrial policy; the formulation and implementation of industrial policy, almost by definition, expands the role of the government and stretches the boundaries within which interest groups operate.

On the other hand, perhaps the most obvious disadvantage of eschewing a conscious and systematic industrial policy is that of losing the flexibility that comes with industry-specific policymaking. America's economy has become so large and diverse that applying uniform policies or relying simply on macroeconomic instruments cannot always be effective in solving specific problems in a sectorally fragmented economy. What helps the semiconductor industry may turn out to be wasteful and inefficient for, say, the textile industry. The old-line industries, high technology, energy, agriculture, and the services constitute five almost separate and distinctive subeconomies, and if only general policy instruments are used, they may prove to be too blunt.

Moreover, the very high costs of backing into a protectionist industrial policy, especially for America's competitive industries such as semiconductors, would very likely result in inflation and allocative inefficiency and trigger an adverse chain of repercussions throughout the economy. In order to bring inflation (exacerbated by protectionism) under control, the Federal Reserve usually has to adopt restrictive monetary policies through the imposition of high interest rates. This in turn tends to slow business investment and overall growth rates, resulting in lower revenue for the government.

Under the current floating system, high interest rates can also cause major shifts in the yen-dollar exchange ratio, as happened in 1982 when the yen weakened sharply in relation to the dollar. An under-valued yen, needless to say, undercuts U.S. price competitiveness and makes Japanese goods more attractive worldwide. Shifts in exchange rate have the effect of widening an already huge U.S.-Japan trade imbalance.[52] For the U.S. semiconductor industry, already hard pressed by Japanese price-cutting in 64K RAMs, an erratic yen-dollar exchange rate badly out of alignment with purchasing power parity is certain to increase the headaches of trying to compete against the Japanese.

A backward-looking industrial policy is also likely to lead to inefficiencies in the allocation of capital and labor, if smokestack industries like steel and autos, which are losing competitiveness but find relief behind government protection, crowd out high-growth industries in their quest for investment capital. Such distortions will rigidify political alignments in support of protectionism and eventually delay the shift in international comparative advantage; this will give rise to serious trade frictions and ultimately will take a toll in productivity rates. If past experience is an accurate guide, protectionism will not restore competitiveness to inefficient smokestack industries; it will merely prolong and deepen its dependence without improving efficiency.

A backward-looking industrial policy can also shift the weight of the effective tax burden away from the declining sectors and onto the shoulders of healthier industries. One study, based on comparative 1973 data, has shown that the coefficient of variation in effective capital tax rates and in the ratio of effective capital tax to labor tax rates in the U.S. is substantially higher than it is in Japan.[53] There is a much wider spread between America's industrial sectors, with politically influential industries paying less than other, often more efficient producers. The pattern reveals little overall coherence, much less the guiding hand of economic logic; it gives the appearance of being makeshift, ad hoc accommodations to political pressures.[54] Widely inequitable tax treatment results in a net transfer of income and tends to penalize sectors like the U.S. semiconductor industry, on which much of the U.S. economy's future growth depends. A protectionist industrial policy therefore hurts America's economy in a variety of ways and systematically discriminates against some of the very sectors on which future productivity hinges.

Of course, no country is completely free of market-distorting protectionism. Japan has its own problems with respect to agriculture. The direct subsidy to farmers and beef producers is massive. Although the transfer of income is politically understandable, it unleashes inflationary pressures and leads to serious economic inefficiencies, just as in the U.S. The difference is that, so far at least, the Japanese have not had to face the problem of coping with a great many declining industries clamoring for protection, adjustment, and relief. It is not clear how Japan's political system will react if and when that problem arises. So far, the most egregious example of inefficiency has been the agricultural sector, and for reasons already explained, the virus of backward-looking protectionism has not yet spread very widely to the industrial sectors. The forward-looking nature of Japan's industrial policy does not discriminate against the semiconductor and other high priority industries; indeed, the mix of measures is intended to do just the opposite—that is, create an environment that systematically fosters their growth. Small wonder the U.S. semiconductor industry feels it has a two-front battle to wage: one with its formidable Japanese competitors and the other with its own government.

For the U.S. semiconductor industry, it must be frustrating and the cause of envy and concern to see Japanese competitors reap the benefits of custom-tailored treatment—selective exemption from antitrust, special tax provisions, subsidies for national research projects—when specialized treatment seems unattainable in America. The asymmetry must be especially galling because the Japanese semiconductor industry seems to narrow America's lead each time the government fine-tunes its industrial policies or gears up for a national project. But in America's federal system there is nothing to prevent state governments from playing a more active role in microindustrial policy. With industries concentrated in certain regions, state governments have an incentive to design policies that promote locally based industries. California, for example, is in a position to encourage the growth of the semiconductor industry though investment-inducing tax measures, allocation of funds for semiconductor research at state universities, changes in the education system aimed at producing a larger number of better-trained engineers and scientists, and so on. Though the federal government is reluctant to resort to microindustrial planning, state and local governments may be capable of fulfilling that role to some degree.

Financing. One fundamental feature of an industrial policy, assuming that it works properly, is to channel capital, manpower, and resources away from declining industries toward high-growth, higher value added sectors. For the semiconductor industry, and for high-growth industries in the past, perhaps the prime benefit of Japan's industrial policy has been the access to capital made available by the prioritization of national needs. During the early years of capital scarcity (the 1950's to the mid-1960's), the state's role in funneling capital to priority industries was critical, not only in terms of subsidies and loans allocated from state coffers but also in terms of clear signals that the substance of industrial policy communicated to private banks. Money supplied from the government's Fiscal Investment and Loan Program (FILP) (*Zaisei Toyushi*) provided a significant portion of funds to industry (see Table 9).

Table 9 shows that nearly 30 percent of all funds for private industry came from FILP sources during the early fifties. For priority industries, such as electrical power generation, shipping, coal, iron, and steel, the figure was nearly 40 percent.[55] Moreover, public funds through government-affiliated banks such as the Japan Development Bank and the Industrial Bank of Japan were usually extended on a preferential basis at interest rates significantly below long-term prime rates charged by private banks. In the early sixties, the interest rate differential was over 3 percent. And the extension of such loans and subsidies served as a "signaling mechanism" to the private banks, which followed the government's lead in funneling funds to the capital-hungry sectors. What government funds and guidance provided, therefore, was a secure supply of capital at preferential rates for priority industries. The government thus laid the foundations for Japan's industrial infrastructure and rapid economic development.

It should be pointed out, however, that, until around the early-to-mid-seventies, the large electronic firms (also the major semiconductor manufacturers) were not among the prime beneficiaries of Japan's allocative system. In the fifties the majority of JDB "lead" loans went to the basic infrastructure industries (electrical power generation and transport), and then to the heavy manufacturing industries (steel, chemicals, and machinery). By the time these industries had built the bulk of their new plant facilities, the conditions of capital scarcity had loosened considerably, and government funding as a proportion of total industrial funding had shrunk to less than half of what it had been in the early fifties. Indeed, from the mid-seventies on, Japan found itself in the unusual position of having an excess of private savings over business investment demand.

Under these circumstances, government-led financing has declined in relative importance. Corporations today are perfectly capable of obtaining capital even without making the list of priority industries. Banks and security houses are eager to do business with sound companies of high growth potential whose profits show a positive upward trend. The JDB still continues to make preferential loans, and the

TABLE 9

Funding Sources for Japanese Industries, 1952-1975 and 1980

(Percent of total funding)

Funding source	1952-55	1955-60	1961-65	1966-70	1971-75	1980
Capital market	11.9%	21.6%	17.8%	11.2%	12.2%	10.5%
Private banks	59.8	60.7	66.4	73.7	74.2	75.8
FILP:	28.3	17.7	15.8	15.1	13.7	13.6
JDB	13.3	4.6	4.2	3.9	3.6	2.3
Others	15.0	13.1	11.6	11.2	10.0	11.3

SOURCE: Yukio Noguchi, "The Government-Business Relationship in Japan," unpublished paper prepared for U.S.-Japan Economic Symposium, Honolulu, January 1981, p. 30.

knowledge-intensive industries have been given favorable treatment: in 1981, the computer, electronics, and machinery industries received a total of over $300 million.[56] But the prominence of JDB loans as "signaling mechanisms" has faded, and the interest rate differential between public and private loans has narrowed; what used to be over 3 percent in the early sixties has dropped to only one percent in 1981.[57]

The transformation of Japan's financial system has led to fundamental changes in the nature and function of industrial policy. Government-guided financing is no longer its dominant feature, nor among the most crucial reasons for its effectiveness. Industrial policy today is preoccupied with adjusting Japan's interdependent industrial economy to constant changes, conflicts, and opportunities in the international environment.

On the other hand, an area of government involvement that has had a substantial impact on the competitiveness of the semiconductor industry is macroeconomic management. If the government is able to sustain steady growth rates while keeping down inflation, interest rates, and unemployment, the direct and indirect benefits to the semiconductor industry will be enormous. Since 1975, Japan's record of macroeconomic management has been unsurpassed by that of any large industrialized state, including the United States. This record is a tribute to the Japanese government's skillful handling of monetary and fiscal policies. But from an American perspective, the record is objectionable in that (1) the lack of multilateral macroeconomic coordination has contributed with disconcerting frequency to serious yen-dollar exchange rate disequilibria and large trade and current account imbalances; and (2) lingering restrictions on foreign access to Japan's capital market have impeded the "self-correcting" mechanism of exchange rate adjustment. Closer macroeconomic coordination and freer access to Japan's capital market could go a long way toward redressing some of the exchange rate burdens currently borne by the U.S. Although there may be domestic reasons for the government's reluctance to open up its financial system fully, its foot-dragging is viewed abroad as evidence of a deep-seated national self-centeredness unbefitting the world's second largest market economy.[58]

Another important but seldom mentioned feature of government financing for priority industries is its relative insulation from the pulling and hauling of partisan politics. This is attributable largely to the power of the economic bureaucracies and close government-business relations, as pointed out earlier; but it is also due to the separation of the Fiscal Investment and Loan Program, which supplies funds for priority industries from the General Accounts Budget (GAB), which allocates money to the labor-intensive sectors. At the risk of simplification, one might say that FILP took the lead role in advancing targeted industries and accelerating economic growth while GAB redistributed income to the slower growing, lower value added sectors, and smoothed out some of the political and social dislocations caused by rapid structural transformation. Because such politically powerful groups as the farmers derived their subsidies from GAB, intense lobbying efforts came to be centered on the GAB allocation processes. By contrast, because FILP did not even require Diet approval until 1974, it was never as vulnerable to interest group lobbying or the LDP's pork barrel intercession. A significant portion of money could be set aside each year for priority industries. Since growth industries like semiconductors have usually not had the time, manpower, or resources to acquire the political pull of older industries, they are generally not in a strong position to compete effectively in the budgetary arena. To a large extent, FILP has obviated that need. The semiconductor industry as a designated priority has been assured of public funds at lower than commercial interest rates, and it has not had to enter the lobbying fray with old-line manufacturing behemoths.

Summary conclusions. We have seen that the efforts of the Japanese and American governments to foster the development of their semiconductor industries have worked well within the needs and objectives of their respective systems. American and Japanese semiconductor industries are far ahead of all foreign competitors, thanks in part to direct and indirect government facilitation, particularly during the early phases of development. Of course, the main reason for success has to be attributed to the dynamism of their private sectors. As much as the government has done, its role has unquestionably been secondary and is best understood in terms of how it has affected the competitiveness of individual companies, the primary driving force behind semiconductor development. Were it otherwise, the centrally planned states, not the market-oriented economies, would be at the cutting edge of technology and industrial leadership.[59] Both Tokyo and Washington, being aware of the importance of market forces, have tried not to tamper excessively in ways that would thwart these forces or dampen the structure of capitalistic incentives.

Of the two, Tokyo has probably done more to help its semiconductor industry than Washington. Yet, even in Japan's case, it would be a mistake to ascribe success primarily to government policies. Some of Japan's best-known firms—Sony, Matsushita, Seiko, Casio, Sharp (producers primarily of consumer electronics)—have grown and prospered without being the prime beneficiaries of preferential government treatment—such as participation in national research projects, public procurements, Japan Development Bank loans and other subsidies. Such policies have tended to benefit a handful of established firms—NEC, Mitsubishi Electric, Toshiba (highly diversified producers of consumer goods and electrical machinery)—with which other electronics firms have been in fierce competition.

The United States still maintains its dominance in the area of high technology, and were it not for Japan, the U.S. system might not be viewed as being in such disarray. For all the preoccupation with the "Japanese threat," the fact is that America's semiconductor industry is still in a commanding position in terms of technology, innovativeness, and world market share. It is not likely to relinquish leadership in the foreseeable future. Japanese companies have made major inroads into narrow segments of product markets, especially the MOS dynamic memory chips and CMOS static random access memories, but their share of other U.S. markets has been limited, and they are not yet a major factor in the technologically more advanced markets for CMOS logic, custom circuits, applied software, and peripheral devices.

The Japanese government has gone about the task of industrial promotion intelligently. But the state's role is only a part of the semiconductor story, and probably not the most important part at that. Japan's "pursuer" system has functioned well enough to bring its semiconductor industry from far behind in a comparatively short period of time. But the real test still lies ahead: it revolves around the question of whether Japan's system is now capable of making the tough transition from pursuer to pioneer.[60]

FIVE

Finance

M. *Therese Flaherty and Hiroyuki Itami*

During the 1970's competition in semiconductor markets was based primarily on new technology, relative costs, marketing, and product reliability. The quality of operating professionals seemed to determine the relative success of companies. In particular, capital requirements did not prohibit entry, and most firms seemed to find the capital to fund projects their managers thought worthwhile. U.S.-based companies dominated the world markets for digital integrated circuits. But in the early eighties the U.S.-based Semiconductor Industry Association sounded an alarm.[1] It warned that the outcome of the emerging global competition in high-volume digital integrated circuits might be determined primarily by the cost and availability of capital to the participants rather than by the strength of their operations.

Two factors combined to make this alarm particularly disturbing. First, the U.S. semiconductor industry was generally perceived as a major contributor to new technology developments and growth in sectors such as telecommunications, computing, consumer electronics, and defense. Second, a major alleged cause of the financial differences was Japanese government policy.

These factors, of course, transcend the semiconductor industry. In an era of increasing global competition in many industries, several questions arise. How do separate and different national financial institutions affect the international competitiveness of indigenous firms? If the differences arose independently during the evolution and growth of the national economies, should they be treated in the same way as differences due to directed government policies? The U.S. government's industrial policy had been limited in principle to establishing the ground rules for free markets: was a substantive industrial policy required in the new order of international trade?

The simple economic theory of comparative advantage suggests that semiconductor producers should cease production of unprofitable semiconductors (whatever the source of cost disadvantage) and move to products in which they have a comparative advantage. But this economic theory is static, and the semiconductor industry is fundamentally dynamic. Its technology and products change rapidly as compared with most industries; the spillover effects from the semiconductor industry continually generate new products and processes in other industries. It is on the basis of future advantages outside their semiconductor industries—not on the basis of current costs—that the governments of France and Japan explicitly justify their support of national semiconductor industries. Indeed, in the early 1980's some economists provided a new basis in economic theory for the assertion that one government acting to subsidize its R&D-intensive industry can give that industry a competitive edge if other governments remain passive.[2]

But the direct evidence that the Japanese government had significantly subsidized the semiconductor manufacturers was not available. In fact, the magnitude of direct government subsidies to semiconductor companies was not large. The case that the Japanese semiconductor manufacturers possessed a financial advantage in international trade rested on indirect Japanese government policies toward the semiconductor industry, on macroeconomic policies, and on the special nature of the Japanese financial system.

It would be impossible in one chapter to compare all aspects of the Japanese and U.S. financial systems and government financial assistance to the respective semiconductor manufacturers. Accordingly, we focus on those differences that are most likely to have significant impacts on the finance of semiconductor operations. The differences that we discuss could, in turn, have a significant impact on the long-term competitive positions of many U.S. and Japan-based semiconductor manufacturers. However, it is important to note that many of these differences were not designed to serve national competitive objectives in the global semiconductor market. Rather, U.S. and Japanese financial institutions evolved to meet the needs of their respective economies. Nor do we intend to suggest that either country remake its financial system in the image of the other. Rather, we believe that there are actions that the governments and companies could undertake separately and in cooperation to attenuate the difficulties that may arise from the differences between their financial systems. We hope in this chapter to advance the general understanding of the financial institutions in the U.S. and Japan and the nature of their impact on the emerging global competition in semiconductors and in other global markets.

In this chapter we first compare the capital formation processes in Japan and the U.S. The macroeconomic policies of both governments played significant roles in determining the degree of access to their own and other sources of corporate capital, the size of those resource pools, and the costs of corporate capital. In the next section we discuss capital allocation among the companies in each country. This section considers the returns and risks on corporate investments. It also contrasts the administrative character of the Japanese financial system with the more market-oriented character of the U.S. system. Several salient features are the important role of the city banks and insurance companies in Japan, the administrative transaction costs of monitoring high-risk borrowers, and the different effects of government attention in an administration system like that in Japan as opposed to a more market-oriented system like that in the U.S. The next section describes the companies in the U.S. and Japan that participated in the semiconductor race, showing individual differences among companies in each of the two countries as well as differences between the two national industrial groups. In the last section we discuss the impact of the different financial institutions on the outcome of the semiconductor race.

Capital Formation

We begin by describing some of the basic differences between the financial institutions in the U.S. and Japan during the 1970's and early 1980's. The first of these was Japan's greater isolation from world sources of capital. The Japanese Ministry of Finance (MOF) historically exercised strict control over access to Japan's financial institutions. Until recently, Japanese sources of capital were effectively off-limits to non-Japanese companies. At the same time, both foreign and Japanese sources of capital were restricted to Japanese companies. MOF regulated the issuance of bond and equity issues, most interest rates, and the amount of money banks could lend. MOF also limited the amount of money a Japanese manufacturing company or financial institution could take from the country.

During the 1970's and early 1980's many Japanese companies seemed to find foreign capital sources very attractive, though very difficult to access. Their beginning investments in plants and equipment in other industrialized countries during that period created new needs for foreign exchange. Further, companies based in Japan wanted to avoid any restrictions MOF might put in place in reaction to an external event. That Japanese companies desired a more stable and consistently available source of capital outside Japan is suggested by the queue of Japan-based companies awaiting MOF approval in 1981 to raise capital in New York and London. When their turns came they offered bonds even though the high level of Western interest rates prohibited U.S. and European offerings. By internationalizing their operations and their finance, these Japan-based companies were, of course, also slowly loosening MOF's control of the Japanese financial system.

The Japanese financial system has been characterized by continual evolution during the postwar period. It would be a serious mistake to suggest that the system of the 1980's is little changed from that of the 1950's. Nevertheless, the Japanese financial institutions continued to be insulated from the international capital market. Beginning in the mid-1970's there was considerable discussion of liberalizing and opening Japan, but free capital movement was not sufficient for arbitrage to bring the two financial systems together effectively.

Certainly, this quick description oversimplifies the Japanese financial system. But the separation of Japanese corporate finance from the rest of the world was generally acknowledged by Japanese bankers and economists. It is a critical basis to our discussion. From a macroeconomic standpoint, the control permitted by this separation allowed MOF to facilitate adjustment to international shocks such as the various oil crises.[3] It also probably helped to channel funds toward the task of building the nation's industrial infrastructure after the Second World War.

The separation of Japanese financial institutions from those of the Western industrialized nations of course meant that the savings of Japanese individuals stayed in Japan. Japanese savings did not go to higher return investments in Europe or Latin America or the U.S.—at least most of the Japanese savings did not. Rather, the savings stayed in Japan to become the pool of external capital from which the government and companies borrowed. By world standards, the supply of capital in Japan was prodigious. During the postwar era, savings as a percent of GNP in Japan was considerably greater than 20 percent while in the Western democracies it was never greater than 20 percent. In 1980, for example, U.S. gross savings totaled 15 percent of GNP. These are, of course, gross figures to which adjustments should be made for precise comparison.

The supply of capital available to the Japanese industrial sector as a whole seems to have been proportionally greater than that available to industries in Western countries. Corroboration of this point comes from a comparison of the reliance of Japanese and U.S. firms on external capital sources (generally loans, equity, and bonds) as opposed to internal sources like retained earnings for investment capital. Japanese firms in general relied 15-20 percent more—and Japanese semiconductor firms as much as 45 percent more—on external financing than their American counterparts. For U.S. semiconductor firms external financing as a percent of total financing averaged 21 percent between 1975 and 1979. For Japanese semiconductor manufacturers it was 45 percent (see Tables 11 and 12). If the same percentages applied to the two industries in 1980, then the U.S. industry would have borrowed $290 million from outside sources and the Japanese industry would have borrowed $315 million. The Japanese would have acquired more external capital than the U.S. even though the U.S. industry invested over twice what the Japanese industry did. (Investment estimates from Tables 16 and 17.)

The third feature of the financial systems we shall examine is the cost of capital, or the price a company must pay in the future to borrow a U.S. dollar or its equivalent in yen. The effective separation of Japanese financial institutions meant that at any point in time the cost of capital in Tokyo could differ from that in London or New York. Of course, measuring and comparing the cost of capital to corporations in two countries is fraught with difficulties. One wants to compare like corporations' costs of capital at the same point in time, but like corporations in different countries do not come neatly paired; also the cost of capital within any one country at any particular time differs among companies owing to perceived differences in the mean and variance of their returns. Although financial theory suggests a way of weighting the costs of equity and bonds to a company at a fixed point in time to arrive at the cost of capital for the company, as well as a way to separate diversifiable from nondiversifiable risk, a company's estimated marginal costs of capital must be adjusted for perceived risk, for "appropriate" inflation rates, and for exchange rates, among other things. A completely satisfying methodology for doing so has not yet been developed.

Nevertheless, several studies have compared capital costs between the U.S. and Japan and found large differences. One of the most sophisticated studies was headed by Hatsopoulos and his associates at the American Business Conferences, Inc.[4] They found that during the early 1980's the cost of capital in Japan was considerably less than that in the U.S. But since the authors had to make numerous assump-

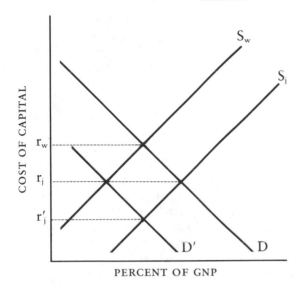

Fig. 5. Costs of capital (r) in Japan and the West in relation to savings (S) and demand (D).

tions in order to arrive at their conclusion, reasonable people could quibble with their findings. Though we would question some details of the study, we agree with the general conclusion that the cost of capital to large corporations in Japan was considerably less than that in the U.S. during the late 1970's and early 1980's. The argument for our concurrence is macroeconomic and is graphically depicted in Figure 5. Since Japan's financial system was effectively isolated during that period, there must be different equilibria in the two markets. In Figure 5, the savings schedule for Japan (S_j) lies to the right of that for the West (S_w) because the rate of savings as a percent of GNP was considerably greater in Japan than it was in the West. Both schedules slope up because we presume that a greater supply of money for investment would be made available by consumers if companies would pay more for it.

The amount of capital demanded by companies increases as the price they pay for it declines. The aggregate demand for capital in an economy is based on opportunities for new investment. Japan has a "dual economy"; it has the general industrial composition of the typical industrialized economy, but it also includes a significant sector of inefficient, less advanced, small businesses in, for example, the agricultural and restaurant sectors. This means that the aggregate demand curve for capital in Japan must on our figure coincide with that for the West (D) or lie below it (D'). Suppose that the aggregate demand curves for capital in the West and in Japan are the same relative to GNP; then the cost of capital in the West is r_w (the intersection of S_w and D), and the cost of capital in Japan is r_j (at the intersection of S_j and D). This illustrates the general economic proposition that when supply increases and demand stays the same, the price in the market decreases. Since the supply of capital in Japan was greater than that in the West, its price was lower in Japan. (Had the demand been less, the cost of capital would have been even lower, r_j.)

Thus, leaving aside considerations of Japanese government loan guarantees and subsidies and directives, it appears that the separation of the Japanese financial system from that of the West and the greater supply of savings in Japan were sufficient to produce lower costs of capital in Japan. Of course, we do not intend to suggest that these two factors alone produced a lower cost of capital. Different institutions to allocate capital among firms, different information systems, and very different systems of intermediation play roles, as do Japanese government directives in providing subsidies, in organizing, and in focusing opinion and providing information. These sufficient conditions—separation and high savings rate—are especially important for the policy analyst. It may be that no condition other than direct subsidy is sufficient to result in a lower cost of capital or a greater supply of it. Thus any other differences between the two financial systems must be considered in light of their separation and relative savings rates.

Capital Allocation

Given the general conditions of cost and amounts of external capital available in Japan and the West, we come to questions about how the financial institutions in the two countries allocate the supply of capital among companies. The many differences in corporate finance and institutions between Japan and the U.S. are widely recognized. For example, the debt/equity ratios of many Japanese companies were higher than those of their counterparts in the U.S. Until the early 1980's there was, in effect, no source of venture capital in Japan whereas there was an active one in the U.S. The thirteen city banks and life insurance companies in Japan played a much larger role in Japanese corporate finance than the largest U.S. institutions. The Ministry of International Trade and Industry (MITI) played a strong role in articulating national industrial policies and coordinating R&D, and was a much more prestigious and persuasive group than any comparable government institution was in the U.S.; and MITI widely argued that certain sectors like

TABLE 10

Indirect Financing in the U.S. and Japan, 1966-1978

(Percent of total external financing)

Period	Indirect Financing		Indirect financing through banks	
	U.S.	Japan	U.S.	Japan
1966-70 average	78.2%	92.0%	31.2%	62.9%
1971-75 average	77.4	91.3	31.9	61.0
1975-78 average	76.0	89.5	27.2	53.3

SOURCE: Shoichi Royama, "A Perspective of Comparison of Financial Institutions," *Contemporary Economics*, Summer 1981 (in Japanese).

electronics and semiconductors ought to be favored by investors for the economy's future.

As with most generalizations, these were partial truths. Reality was more varied. And in the 1980's that reality seemed to be changing and converging in many ways.

Extent of intermediation. We observed above that in Japan external financing (as opposed to internal or retained earnings) was larger in absolute magnitude and much larger in proportion to total investment than in the U.S. The general category of external financing, however, should be split into two: *direct external financing* (through bonds and equity issues) and *indirect external financing* (through loans from banks and other financial intermediaries). In connection with the high debt/equity ratios and the importance of the banks in Japan, it was probably the role of indirect external financing that was most discussed during the early 1980's.

In the period 1975-78, nearly 90 percent of the external financing of all Japanese companies was indirect, or channeled through financial intermediaries like city banks and life insurance companies. Table 10 shows that this was 13.5 percent greater than the portion of indirect external finance for all companies in the U.S. More than half of all external financing in Japan flowed to companies through the city banks. In the U.S., banks controlled only slightly over one-fourth of the outside capital acquired by companies each year. In the Japanese economy as a whole, the city banks' share of indirect financing decreased during the late 1970's as the money and bond markets increased in size and

TABLE 11

Sources of Finance for U.S.-Based Semiconductor Firms, 1970-1979

(Percent of total funds financed)

Period	Internal funds	Equity issues	Bond issues	Loans[a]
1970-74[b] average	61.8%	19.6%	9.4%	9.2%
1975-79[c] average	79.0	6.6	8.6	5.8
1970-79 average	70.5	13.0	9.0	7.5

SOURCE: Annual reports of firms.

   NOTE: Data are averages of six of the firms listed in Table 16; for AMD and Signetics, financial data were not available for several years in the 1975-79 period because of mergers.

   [a]Includes both long-term loans from various financial institutions and net increase of short-term loans.

   [b]For some firms data for 1970 or 1971 are missing.

   [c]For the acquired firms (Fairchild and Mostek), 1979 data are missing.

became more accessible, but even at the end of the period, the share of the banks in Japan was much greater than that of U.S. banks.

This pattern shows even more clearly in the financial reports of the semiconductor manufacturers. Here we refer generally to the high-volume digital integrated circuit manufacturers listed below in Table 16 for the U.S. and in Table 17 for Japan. The companies differed in many ways, some of which are described below. In interpreting these tables it is important to note that financing is defined as total financing to the parent company. The Japanese companies that made semiconductors were all highly diversified electronics companies. Several U.S. companies like Motorola and Texas Instruments also had significant non-semiconductor operations. All these companies consolidated their operations in obtaining external funding. Consequently, data on the companies' total external financing—not on external financing applied to semiconductor operations—are used. Moreover, internal financing includes earnings retained from all operations.

Tables 11 and 12 present indicators of the sources of finance for the U.S. and Japanese companies during the 1970's. In comparing the data for two countries, one should note that in the first five-year period (1970-74) many of the American firms were in their formative stage and depended substantially on equity issues as the initial capital. Therefore, 1975-79 data are more informative of financing patterns of ongoing American firms. It is also worthy of note that long-term loans in these tables are counted as gross new loans. For short-term loans, only the net increase is included. When there was a net decrease of

TABLE 12

Sources of Finance for Japan-Based Semiconductor Firms, 1970-1979
(Percent of total funds financed)

Period	Internal funds	Equity issues	Bond issues	Loans[a]
1970-74 average	47.2%	2.9%	8.8%	41.1%
1975-79 average	55.2	7.3	10.7	26.8
1970-79 average	51.2	5.1	9.7	34.0

SOURCE: Annual reports of firms.

NOTE: Data are averages of seven of the firms listed in Table 17, excepting Tokyo Sanyo and Oki Electric, the two smallest. Tokyo Sanyo is a subsidiary of Sanyo Electric, and Oki had a major financial difficulty during this period; including them in the sample would tend to distort the average Japanese financing pattern.

[a]Includes both long-term loans from various financial institutions and net increases of short-term loans.

short-term loans outstanding, it is considered as an application of funds. Thus short-term loans may increase, while long-term loans outstanding decrease as Japanese firms borrow money to repay their old loans. Finally, the data apply not only to semiconductor operations but to the entire firm. For the diversified electronic firms in our sample, it is impossible to know which funds were applied to finance semiconductor operations.

The most striking difference between the two countries in the 1975-79 period was the magnitude of loan, or indirect external financing. Japanese firms depended much more heavily on the loans (mainly from the banks). The share of loans did decrease from 41 percent in the early 1970's to 26.8 percent in the late 1970's. The fact that Japanese firms relied on the loans only for a quarter of their total funds needs may be a little surprising. Internal funds comprised, of course, the largest source of funds for Japanese firms.

The differences in loan financing just about balanced the difference in internal financing. In other words, the American firms provided 24 percent more of their financing needs from internal sources than the Japanese firms did. The Japanese firms received about the difference in percentage terms (21 percent on average) of their total funds from external sources in the form of loans. This difference suggests a considerably greater inflow of capital to Japanese semiconductor firms than to their U.S. counterparts.

The shares of equity financing and bond financing were just about the same in the two countries. Equity issues played a very minor role in both the U.S. and Japan, as a percentage of total funds (around 7 percent); but among the external financing sources, equity financing played a more important role for the U.S. semiconductor firms (about 31 percent of externally financed funds) than for their Japanese counterparts (about 16 percent of external funds). Loan financing was about 60 percent of external financing for the Japanese firms.

As mentioned above, a part of Japanese loan financing was used for repayment of old loans, thus leaving the loan outstanding at a relatively stable level. In fact, for 1976, 1977, and 1978, many Japanese firms decreased their loans outstanding, although their gross new loans were at a positive level. This was the case for Hitachi, Toshiba, Mitsubishi, Matsushita, and Sharp. Fujitsu started decreasing its loans outstanding in 1978 and continued doing so in 1979. NEC's loans outstanding decreased in 1978 but showed a small increase again in 1979. Before 1975, most Japanese firms continually increased their loans to finance their growth.

Between the early seventies (1970-74) and the late seventies (1975-79), Japanese firms seem to have changed their financing strategy by reducing their dependence on loans and relying more on internal funds and equity financing. It is difficult to discern any major changes of American financing strategy from Table 11. As mentioned above, the greater dependence on equity in 1970-74 was largely due to the initial capital formation of firms like Intel, Mostek, and National Semiconductor. There may have been a trend during the early 1980's toward more debt financing, especially bond financing, among the American firms to cope with rapidly rising capital requirements. Both Texas Instruments and Intel issued convertible bonds in 1980 ($200 million for TI and $300 million for Intel), breaking away from their tradition of almost complete internal financing, but this may have been a response to the stock market cycle rather than a long-term trend. In fact, in late 1982 IBM agreed to acquire 12 percent of Intel's equity and an option to increase its holdings up to 30 percent.

It has been suggested that at least part of the Japanese loans was more a disguised form of equity investment by the financial institutions than outright loans.[5] According to this line of interpretation, some of the Japanese city bank loans were very much like preferred stock from the standpoint of the borrowing firms: the borrowers paid interest on preferred stock before distributing their earnings to common stockholders, and the principal often was not to be repaid to the banks because the loans were routinely renewed. Following this suggestion, Table 13 recomputes Table 12 on the assumption that part of

TABLE 13

Adjusted Sources of Finance for Japan-Based Semiconductor Firms,
1970-1979

(Percent of "real" total funds)

Period	Internal funds	Equity issues	Bond issues	Adjusted loans[a]
1970-74 average	52.8%	3.4%	10.0%	33.8%
1975-79 average	62.8	8.0	11.7	17.5
1970-79 average	57.8	5.7	10.9	25.6

SOURCE: All data as in Table 12.
   [a] Adjusted loans equal actual loan financing (as in Table 12) minus 1970-71 average of repayment of long-term loans.

new long-term loans were loans made to repay old ones. In Table 13, we arbitrarily assumed that in each year every Japanese firm renewed long-term loans equivalent to the average repayment of long-term loans for 1970-71. Thus each firm's two-year average repayment (1970 and 1971) is excluded from the total funds to yield figures adjusted for "real" financing need. In other words, they are treated like an equity investment made much earlier.

The evidence of Table 13 is, as one might expect, somewhat stronger than that in Table 12 in support of the same propositions. The Japanese semiconductor firms reduced their reliance on indirect and direct external financing during the 1980's. Whether the corporate finance structure of Japanese firms would continue to move toward that of U.S. firms was, of course, an important question for international trade.

Tables 11 and 12 also show that on average loan financing for semiconductor companies between 1975 and 1979 accounted for 60 percent of external financing in Japan and 28 percent in the U.S. For the entire period, the disparity was even greater: 25 percent of external financing was in the form of loans for the U.S. companies while 70 percent was in loan form for the Japanese companies. These differences are somewhat attenuated in the adjustments made for Table 13, but the general conclusion remains: indirect financing in the form of bank loans played a much larger role in external financing of Japanese semiconductor companies than of their U.S. counterparts.

Not only were banks in Japan suppliers of adjusted loan capital, they were also major suppliers of equity capital. United States law prohibits banks from owning the equity of their borrowers. But Japanese banks are in many cases major stockholders and bondholders of their client firms. (There were, however, ceilings on the banks' equity investment in their customers, and the ceiling was lowered in 1983 from 10 percent to 5 percent.) For example, Table 14 presents the holdings of the ten largest stockholders of NEC as of March 1980. There are three banks in this list, Sumitomo Bank, Sumitomo Trust and Banking, and Mitsubishi Trust and Banking. Sumitomo Bank was the largest supplier of short-term (but renewable) loans, and Sumitomo Trust and Banking was the largest supplier of long-term loans to NEC. Sumitomo Bank was also NEC's largest creditor, lending $188 million of short-term and long-term capital as of March 1980 (roughly 17 percent of NEC's total loans outstanding).

The six banks and insurance firms listed in Table 14 (excluding Sumitomo Marine and Fire Insurance) provided 36.8 percent of long-term loans and 35.1 percent of total loans (both long-term and short-term) in 1979. NEC's example is the rule rather than the exception among large Japanese companies. Banks and insurance firms are likewise both major lenders and major stockholders of most other Japanese semiconductor manufacturers.

It is worth mentioning that NEC is a member of the Sumitomo keiretsu, or group, although the significance of this relationship is not clear. The benefits of keiretsu membership were difficult for Caves and Uekusa to identify in their study of groups of companies in Japan.[6] Many Japanese managers we spoke with maintained that by the late 1970's and early 1980's the main effects of group membership coincided with the financing ties. Indeed, companies like Sony that were not affiliated with any group had much lower shares of direct and indirect external finance than the large group-affiliated companies like NEC. Like Caves and Uekusa, we found it difficult to discern financial advantages from group membership that were not coextensive with finance flows.

The fact that indirect financing played a larger role in Japan than in the U.S. immediately suggests that the role of direct external financing (mainly equity issues and bond issues) was smaller than it was in the U.S. This supposition seems to be borne out: in 1975, the holding of marketable securities (stocks, bonds, commercial papers) accounted for 35.9 percent of total nonmonetary financial assets in the U.S.; in Japan, in 1977, the comparable figure was 12.9 percent. In 1975, corporate stocks constituted only 3.1 percent of Japan's nonmonetary financial assets.

TABLE 14

Largest Shareholders of NEC, March 1980

Shareholder	Percent of NEC shares owned
Sumitomo Life Insurance	8.6%
Sumitomo Bank	6.2
Sumitomo Marine and Fire Insurance	3.5
Daiichi Life Insurance	3.4
Nippon Life Insurance	3.3
Sumitomo Electric Industry	3.1
Sumitomo Trust and Banking	2.8
Sumitomo Trading Company	2.8
Mitsubishi Trust and Banking	2.5
Daiwa Securities	2.3
TOTAL	38.5%

SOURCE: NEC Annual Report for 1980.

The greater role of indirect external sources in corporate finance in Japan as compared with the U.S. indicates that administrative—as opposed to market—mechanisms allocated corporate capital to a far larger extent in Japan. One might wonder whether the administrative mechanism in Japan allocated capital to the same sorts of companies and industrial sectors that the U.S. financial system did. One might also wonder whether the transactions cost of the Japanese administrative system was larger than that of running a more market-oriented system like that in the U.S. These questions as well as the nature of the Japanese administrative system are explored below.

Risks of investments in Japan. In the 1970's, compared with U.S.-based companies, large Japanese companies depended much more heavily on debt capital, especially bank loans. The Japanese semiconductor companies were no exception. Many American observers are surprised by the extent of dependence on bank loans and the fact that Japanese banks lent so much to companies with high debt/equity ratios. The average debt/equity ratios for firms in Tables 11 and 12 were, in 1978, 1.42 for Japan and 0.28 for the U.S. All the Japanese semiconductor companies had much higher debt/equity ratios than their American counterparts. Most U.S. semiconductor companies could not have borrowed external capital in any form if they had had debt/equity ratios approaching those of the Japanese semiconductor manufacturers. Smaller Japanese companies, however, were less leveraged than the privileged large companies. Sony, for example, grew up largely without the support of the city banks and funded its growth to a much greater extent than the large companies with retained earnings.

A further difference between the distributions of capital among companies in the U.S. and Japan is seen in venture capital. During the 1970's and early 1980's there was in effect no city bank capital available for new ventures in Japan. In the U.S., venture capital was of course essential in establishing many of the independent semiconductor manufacturers, and the venture capital industry was still growing at a steady pace during the late seventies and early eighties.

These phenomena in Japan are due in part to the administrative mechanism for capital allocation along with macroeconomic conditions. Several plausible connections between the financial system and the system of capital allocation should be examined in detail.

One explanation (which had currency in the U.S.) of the extent of highly leveraged firms in Japan was implicit government loan guarantees. The existence of such loan guarantees proved difficult to substantiate. Many Japanese bankers believe that the government did give some implicit loan guarantees, specifically for projects in which government banks participated.[7] But of the largest lenders to NEC, the most highly indebted of the semiconductor companies (Table 15), the only government bank, the Japan Export and Import Bank, would only have supplied trade credit to NEC's customers abroad—not funds to invest in fixed assets or R&D. So it seems that even heavily in debt, NEC did not depend significantly on government loan guarantees. This may suggest—though it certainly does not prove—that Japanese government loan guarantees through the usual channels did not increase loans to the semiconductor manufacturers during the late seventies and early eighties.

Beyond this, there seem to have been conditions in Japan's postwar economy that made it worthwhile for the city banks to lend to highly leveraged large companies. The city banks seem to have developed close relations with their highly leveraged customers that allowed them to evaluate ongoing risks better and to change management when problems arose within the highly leveraged big companies. But that does not explain why the city banks concentrated their loans to a group of large companies that became highly leveraged rather than distributing the loans over a larger group of less highly leveraged companies.

One explanation consistent with the close relations observed between the leveraged Japanese companies and the city banks has to do with transactions costs and risk. In general, the price a creditor de-

TABLE 15

Largest Lenders to NEC, March 1980

Lender	Percent of total loans
Sumitomo Bank	16.7%
Sumitomo Trust and Banking	11.4
Japan Export and Import Bank	6.7
Yokohama Bank	4.9
Sumitomo Life Insurance	4.6
Industrial Bank of Japan	4.3
Long-term Credit Bank of Japan	4.3
Mitsubishi Bank	4.0
Kyowa Bank	3.9
Japan Bond and Credit Bank	2.3
TOTAL	63.1%

SOURCE: NEC Annual Report for 1980.

mands in return for lending money depends on the risk the *creditor perceives*, and not only on the technical and product market determinants of the borrower's future profit distribution. A creditor—such as a city bank—would incur transactions costs in learning about the technical and market determinants of the borrower's future profit distribution. If the information lowered the perceived risk, the creditor could lend at a lower price. Close relations therefore served to lower risk and make more loans possible.

This analysis assumes the existence of the large pool of savings available for investment, which could be supplied only—or mainly— by the city banks in the form of loans. The equity and bond markets that provided alternatives to bank loans and intermediation in the U.S. were restricted by the MOF in Japan. Companies had no mechanism besides the city banks by which to acquire savings from individuals. So from the company standpoint the corporate finance choice was, in effect, between being leveraged or doing without external capital. In this context, high debt/equity ratios do not indicate the same riskiness that they do in the U.S.

From the standpoint of the city banks, little investment opportunity was possible under the rules of MOF other than lending to companies. The MOF controlled many interest rates, but since there were compensatory balances on loans required by the city banks, real rates could fluctuate.

In Japan the lead bank invested to learn about the distribution of future profits of its lenders. Then other banks in the syndicate relied on the lead bank to prevent losses on the loans. Therefore only one bank needed to incur the transactions cost to reduce risk. Moreover, the banks pooled their risks over many loans, thus reducing individual risks. The lead bank would continually monitor the leveraged company closely depending on how risky it was perceived to be. If the lead bank believed that the borrower had managerial problems, it would do what it thought necessary to improve management policies, sometimes even replacing the management.

Consider again NEC, the most heavily debted firm in our Japanese sample. Table 15 displays the largest lenders to NEC (in NEC's keiretsu, or group). In this case, Sumitomo Bank operated as the lead bank for the entire bank group. This would have been the bank to shoulder the major financial burden as well as the responsibility of monitoring the company's operations for the bank consortium. It also would effect management shakeups and policy changes in the company if it considered them necessary. Though NEC's total loans were spread among a number of banks and insurance firms and came partly from outside the Sumitomo Group (4 percent from Mitsubishi Bank, another keiretsu entirely), Sumitomo Bank's ability to monitor NEC's operations and its future prospects as well as its ability to step into the management of the firm if necessary was critical to handling the large risks.

The city banks generally required disclosure of detailed information on the firm's operation and tried to supplement this by encouraging the bank's officers, not only the top level but also junior level, to maintain close and long-term informal personal relationships with the people of the involved firms. This seems on the surface to be different from the arm's length relationship often found between American banks and firms. This difference was probably even greater between the U.S. and Japanese semiconductor firms than between the entire populations of firms in the two countries because the U.S. high-volume semiconductor companies used significantly less indirect external capital than U.S. firms in general.

But a close relationship between bank and client could be found in the U.S. between banks and the companies for which they supply large portions of the capital. On the other hand, the relation between large, financially stable U.S.-based corporations and their bankers would usually be arm's length. The merchant semiconductor companies raised only a very small proportion of external capital from indirect sources so most of them did not have the possibility of intimate relations with large sources of capital. We suggest that the close and informed relation is used in the U.S. and Japan when the loan is risky, but that relation is more distant when the loan is less risky. Thus the relation between a highly leveraged firm like NEC and its lead bank would usually have been close during periods of danger as perceived by the lead bank and more arm's length during periods when the lead bank was not worried.

This administrative mechanism explains why it might have been rational for city banks to lend to highly leveraged companies and how the city banks mitigated what would appear to U.S. investors to have been very large risks. But it does not explain the bias toward lending large amounts only to large companies.

This, too, can be explained by appeal to transactions costs. Information about borrowers is expensive to gather and it generally reduces the risk city banks perceive about the future profits of their customers. The cost of monitoring a large firm was probably not much greater than the cost of monitoring a smaller firm, but the city banks' benefits from lowering the perceived risk associated with a large firm were probably much greater for large than for small, even though the loans demanded by large firms would be greater. Thus, in an administrative financial system like that in Japan there would be a natural bias for city banks to make more risky loans to a group of large firms.

Unlike Japanese banks, U.S. banks did not control a major portion of the nation's flow of funds. Not only were they prohibited by law from holding equity in the firms they lent to, but also legal restraints made it almost impossible for banks to interfere with the management of failing firms until the firms actually declared bankruptcy. In the Memorex case, for example, the Bank of America had to sit idly by until the last minute, officials there say, watching Memorex fall apart. The American financial system thus promoted an arm's length relationship between banks and industry.

In Japan, the relationships among the city banks and the two-way relations between firms and their lead banks amounted to a virtual "banking-industrial complex" analogous to the U.S. "military-industrial complex." The dealings between the banks and the firms in this complex were not arm's length transactions. The high level of bank borrowing by the Japanese semiconductor manufacturers meant that information flowed with the loans at all levels of the organizations. The transactions were "quasi internal." In the terms of modern economics of the organization, the banking-industrial complex was a governance system shaping the actions of companies and banks. The

Japanese banks were willing to lend and the Japanese firms were willing to borrow so much money, not because of Japan, Inc., and government loan guarantees, but because they had consistent, long-term mutual interests, large amounts of bank-controlled capital, and the power and information to protect them. Japanese firms borrowed so much at the loss of some independence from the bank, more information disclosures, and the two-way demands of a long-term relationship. Sometimes this meant that firms borrowed even when they had no real investment needs.

The closer long-term relations between the banks and the large firms in Japan probably increased the effectiveness of the Japanese financial system's support of the semiconductor industry. The banks, for example, made loans available to the large and profitable firms, like those with which they had established loans in the semiconductor industry, even when credit was tight. Just so long as the city banks believed that the highly leveraged semiconductor firms would profit from international competition in the long run the banks would continue to make finance available to them. Just as important, the banks would make loans available to these firms even during tight money conditions when the banks cut back their credit to other firms.

We should note that this system of two-way relationships in Japan did not prevent serious competition among city banks during the 1970's and early 1980's. During the late seventies many large companies carried large debts but were not at risk operationally, and the Japanese bond and stock markets began to be more accessible to borrowing companies and investors. The city banks competed vigorously for business, and several large companies changed their lead banks. During the same period many large Japanese companies were developing international operations with significant overseas investments. The foreign exchange laws were beginning to be liberalized. The control of interest rates and capital availability by MOF was beginning to weaken. Thus the dominance of the city banks, and the prevalence of highly leveraged large firms in Japan might not survive the long-term trend of freer access to markets. This was definitely a slow change just beginning during the late seventies.

Venture capital in the U.S. In the 1970's and early 1980's, not only did Japan have virtually no venture capital market such as existed in the U.S., but furthermore the Japanese banking-industrial complex made it almost impossible for venturesome entrepreneurs to obtain seed money in the form of bank loans. Not surprisingly there were very few, if any, venture businesses even in the high technology area in Japan.

The main reason was that the Japanese equity market was not well developed to enable successful venture businesses (and their capitalists) to sell their stocks to realize capital gains. Whereas in the U.S., some 298 firms went public in NASDAQ in 1980, that same year there were only 14 new public equity offerings in the Japanese over-the-counter market. In another market for public offerings, AMEX in the U.S. and the Second Section of Tokyo Stock Exchange in Japan, the difference in 1980 was less dramatic but still very substantial: 65 new entrants in the U.S. and 12 in Japan. This difference was not a temporary anomaly. From 1976 through 1980, the total numbers of new public offerings were 600 in NASDAQ and 33 in Japan.

Japanese requirements for going public, set by both the government and the securities industry, were far more stringent than U.S. market standards. Among other requirements, the firm had to be larger than a certain level, and it had to submit dividend and profitability records for many years. This stringency, along with the fact that the Japanese equity market did not absorb as large a proportion of the nation's financial assets to begin with, meant that very few venture stocks were offered on the market.

A second reason for the lack of venture capital in Japan was again related to the limited supply of capital for risks not intermediated by the city banks. There was no special tax incentive for the affluent to provide venture money like the tax shelters allowed in the U.S. tax code. Moreover, capital gains were treated as a part of regular income and regular income tax rates applied for both corporations and individuals, whereas in the U.S. the maximum rate was reduced, in effect, to 28 percent. For a small amount of *public* trading (trading in the stock exchange) of securities, a capital gain was not taxable in Japan for personal income tax. But, as we saw above, it is very difficult for the venture to go public in Japan, and the city banks lend to large firms rather than small ones.

Third, the nature of the labor market limited the supply of entrepreneurs who demand venture capital. Low labor mobility in Japan, at least among top-flight engineers, meant that an entrepreneur had a very small chance of returning to any established or well-regarded firm if he failed in a spinoff attempt. This high cost of failure naturally had the effect of discouraging venture attempts.

Finally, a cultural factor might have been working against a labor supply of entrepreneurs in Japan. Belonging to an established organization seemed to be valued more highly in Japan, whereas independence seemed to be more highly valued in the U.S. In contrast, in the U.S., tax incentives and well-developed equity markets encouraged the supply of entrepreneurs from large organizations, as the Silicon Valley phenomenon seemed to demonstrate. Since the supply of capital and people was readily available, it is not surprising that there were active venture capital markets.

In the absence of a strong venture capital market, semiconductor operations in Japan grew under the umbrella of electronics firms. Whether this lack of entrepreneurs in Japan's semiconductor industry has affected the present or future innovative ability of the industry is an important but unanswerable question. In the early stages of industry evolution in Japan, the electronics firms subsidized their semiconductor operations with their non-IC funds. When they got their bank loans, they included the semiconductor finance needs with those of the rest of the corporation. Both IBM and AT&T financed semiconductor operations in the same manner as their other activities. The Japanese semiconductor operations grew only in those companies that could finance internally or externally a long gestation period. They were large, diversified electronics companies that also used semiconductor technology for some product lines that were already established.

Governments' roles in financing. The federal government is widely believed to have played a major role in financing semiconductor companies in the United States during the 1950's and early 1960's, the early stage of semiconductor technology and the industry. Military procurement at that time provided large funding for the development of product technology and for manufacturing process engineering. During most of the 1960's, of course, the U.S.-based merchant semiconductor companies responded primarily to the requirements of the vast industrial demand by computer end-users. There were companies pursuing government contracts during that period also, but they did not seem to advance the technology as government contractors had done in the earlier days.

During the 1980's the U.S. military was, with considerable fanfare, beginning another large-scale procurement program involving large R&D efforts. The VHSIC (Very High Speed Integrated Circuit) program, however, was greeted with mixed enthusiasm by the merchant semiconductor companies in the U.S. Some contended that the technology being required by the military would have limited commercial application. Radiation hardness, for example, was much less important to computer and television end-users than to missile end-users. The managers of specialized semiconductor companies argued that the program would divert the small number of engineers in the U.S. from commercial to military projects, thus injuring the technological competitiveness of the U.S.-based commercial sector. Other companies, like Texas Instruments and Motorola, were participants in the VHSIC program.

During the early 1980's it was impossible to gauge the technical impact of the VHSIC program on semiconductor technology for the commercial sector. It did seem clear that the program was not providing significant capital or R&D subsidies to the commercially oriented operations, even for the participating companies.

The Japanese government's role in semiconductor financing has been mainly indirect, except for the recent VLSI program. The government has made some loans to the semiconductor industry through the Japan Development Bank at preferential interest rates (one percent below prime rate), but only on a small scale. The total was less than $30 million for the entire industry in 1980. On the other hand, the impact was probably greater than $30 million because loans from the JDB were perceived by the Japanese banking community as signals that the Japanese government would ensure the returns on bank loans made for that project. This sort of implicit guarantee, however, was not a blanket guarantee for all the semiconductor operations of a company, but only a loan guarantee for the project in which the government bank participated. It may not have had much impact on the Japanese semiconductor manufacturers, which were for the most part strong companies that already had solid, long-term relations with the banks.

A major indirect role was played by MITI in selecting the semiconductor industry as one of its target industries. In doing so, the Japanese government did not exercise any authority to direct the lending decision of the banks in favor of the semiconductor industry, but it did gather information and help to form a consensus among the semiconductor firms and financial community in Japan. The very articulation of a rational industrial policy to foster a certain industry undoubtedly increased the attractiveness of that industry as a lending target, even though there was no general government loan guarantee (either explicit or implicit). The banks would channel more capital to that industry only as long as the banks agreed with the government's evaluations of the industry's future prospects.

MITI's announcement of a target industry and its estimates (or goals) for that industry may have had important effects, however. It gave a reliable and well-informed estimate of the industry's future. The estimates went to all the banks and insurance companies so that each institution knew that all the others would also be induced to favor the targeted industry with funds. This information could be particularly persuasive and effective because the number of decision makers in the capital allocation process was relatively small and the funds they controlled were so large. There were only thirteen city banks and three long-term credit banks in Japan. Insurance companies were also fund providers and small in number. As we can see from Table 15, of the largest lenders to NEC in 1980, ten banks and insurance companies controlled more than 60 percent of NEC's total loans.

This method of mobilization of capital through governmental persuasion would have been much more difficult in the U.S. for several reasons. First, there was a consensus that governmental persuasion or intervention was inappropriate and frequently misinformed. Second, external capital flowed directly to the companies from a very large number of individual investors and savers in relatively small bundles. The government would not have been able to contact that many investors at acceptable cost. Sheer numbers simply made persuasion difficult. Third, the well-developed equity and bond markets in the U.S. provided by their pricing mechanisms relatively independent indicators of a company's prospects.

Indirectly, MITI also provided capital to the semiconductor industry in Japan through the funds it provided directly to Japanese computer makers. Minoru Inaba stated that it appears that "the Japanese computer industry still really can't do without Japan Computer Co. Ltd. (JECC)—a vehicle for governmental financial, tax and moral support—despite the industry's appearance of competitive strength."[8] JECC was a rental company established in the early 1960's, following MITI's administrative guidance, by seven companies: Fujitsu, Toshiba, Mitsubishi, Oki, Hitachi, and an NEC-Toshiba sales company. JECC purchased computers from the seven manufacturers and rented them to Japanese clients, relieving the manufacturers of a large investment and a large risk in the rental market. It could borrow from the Japan Development Bank, and it received both a lower than prime interest rate for those loans and an implicit guarantee of its loans from city banks. Its leasing agreements were also more generous to customers than those of IBM and other computer manufacturers who lease directly to customers.

Many observers agree that JECC greatly facilitated the growth in sales of the Japanese computer makers during the 1970's. They also agree that the JECC participants received substantial subsidies for their computer operations. It seems that, as they gained competence, the computer companies relied less on the JECC and sold more computers. In fact, Hitachi withdrew from the JECC for nongovernment procurement in 1972.

To the extent that the vertically integrated Japanese companies used internal funds (or retained earnings) to finance capital expenditures for their semiconductor operations or purchased domestically produced semiconductors, this also amounted to a subsidy for the semiconductor operations. The subsidy that actually got passed on to the semiconductor operations probably coincided with the procurement of semiconductors by the Japanese computer manufacturers—that is, the development of digital integrated circuits in Japan was subsidized by having a larger domestic computer market than would otherwise have existed.

The Japanese government also seems to have indirectly subsidized semiconductor developments during the 1970's through Nippon Telegraph and Telephone (NTT) procurement. NTT was a government-owned, but not government-run, corporation somewhat similar in status to the Japan Development Bank. NTT differed from AT&T, however, in that it did not own its suppliers, it did not issue stock, and its employees had no right to strike. NTT worked closely with four Japanese electronics companies just as (before divestiture) AT&T worked closely with Western Electric. NTT provided technical and financial assistance to NEC, Hitachi, Fujitsu, and Oki in their development of semiconductors for telecommunications and perhaps for computers also. This may well have had substantial impact on the abilities of these companies to enter semiconductor markets.

In fact, when Oki had business troubles during the 1970's, NTT interceded. Oki's president in 1982 was from NTT, and NTT gave Oki financial as well as managerial assistance for its recovery. An outside observer might conclude that saving Oki was a cost without real benefit that would not have been incurred under the U.S. business system. In any case, the impact of the move on international semiconductor competition will probably not be very great during the next five years.

There may well have been more technological spillover between semiconductor development for telecommunications and computers at these Japanese companies than there has been thus far between AT&T and the U.S. merchant semiconductor manufacturers. To the extent that this was true, there was probably indirect government subsidy of the Japanese semiconductor business through NTT. Semiconductor technology for telecommunications and data processing apparently were in the early 1980's becoming more closely related (see Chapters 2 and 3). The possibility for subsidy through the regulated telephone utilities may be similar in the two countries, depending on the details of the AT&T restructuring.

Summary. In Japan, the semiconductor companies had two major sources of financing: (1) internal funds, part of which could be from the non-IC businesses of the diversified operations in electronics; and (2) bank loans, which may be regarded as quasi internal because of the very close and long-term relationship between companies and banks. We may also call the union of the companies and the banks the banking-industrial complex. Internal and quasi-internal financing within this complex occupied a similar role and weight to American internal financing. The fact that the companies and the banks were closely related to each other as one complex was characteristic of Japan.

In contrast, American companies had no "quasi-internal" or "quasi-market" financing in any substantial way. They relied mostly on their internal funds, most of which were IC-related internal funds. External capital was allocated among companies primarily by market mechanisms, and a great many individuals invested directly in the companies' securities, such as equity and bonds. This was in contrast to the Japanese external capital allocation mechanisms by which external capital first flowed from individual savers to a few banks and insurance firms and then was allocated by their administrative decisions among various industrial firms. In a sense, Japanese investors made investments in the banking-industrial complex, whereas American investors made investments in the industrial firms themselves.

At the risk of oversimplification, we may characterize the American financial system as a market-oriented system. The Japanese system had more administrative or organizational aspects to its capital allocation mechanism, and in some cases the line of demarcation between "the company" and "the market" was rather blurred. The quasi-internal capital transactions within the banking-industrial complex were a typical example. In this sense, the Japanese system was an administration-oriented system in that although capital allocation was done both in the market and within the companies, the allocation mechanism had more elements of administrative allocation than the more arm's length process of capital allocation in the U.S. semiconductor industry.[9]

Japanese companies across the board used external funds for a greater proportion of their new investment than U.S. companies do. This may well have been related to the much greater pool of savings available in what was until the early 1980's an essentially closed Japanese economy. Japanese companies probably drew more capital from savings than U.S. companies could have at comparable costs.

Another essential difference between the two institutions was the magnitude of external capital they allocated to their semiconductor companies. Recall that 45 percent of the combined capital expenditures of the Japanese electronics companies come from funds generated from the outside—in contrast to only 21 percent of the combined capital expenditures of the U.S. semiconductor companies. (There is, of course, some disparity here between the more diverse activities of Japanese electronics companies and the more narrowly focused U.S. semiconductor companies.)

Semiconductor Companies

The U.S. and Japanese companies that made semiconductor devices differed from each other in several ways that might have affected their abilities to raise capital or their abilities to compete internationally. As Masuda and Steinmueller point out, the companies differed in size, vertical integration, and diversification.[10] There were also differences among the companies in their sources of capital. In this section we discuss these characteristics and the categorization of companies that they suggest.

A first difference is in internal consumption of integrated circuits by companies active in the high-volume memory and integrated circuit markets. Masuda and Steinmueller note that the Japanese companies participating in the high-volume integrated circuit markets in the late 1970's consumed between 16 percent (NEC) and 44 percent (Oki Electric) of their own integrated circuit production.[11] Matsushita and Sony, which consumed 57 percent and 80 percent, respectively, of their integrated circuit production, focused their competitive efforts on consumer products that incorporate integrated circuits rather than on the international merchant market for integrated circuits. Again, Masuda and Steinmueller contrasted the significant internal consumption of integrated circuits by the major Japanese competitors in the high-volume market with the relatively small internal consumption of integrated circuits by their U.S. merchant competitors in the high-volume memory and integrated circuit markets.[12] IBM and AT&T, on the other hand, consumed 100 percent of their integrated circuit production and did not sell in the merchant market.

There seems also to have been a difference between the U.S. and Japanese semiconductor companies as groups according to the share of their total sales accounted for by semiconductors. As a group, the sales of high-volume advanced integrated circuits accounted for a larger portion of total sales of most U.S.-based companies than of those based in Japan. This, of course, reflected in part the greater internal consumption of the Japanese companies, but it also reflected the

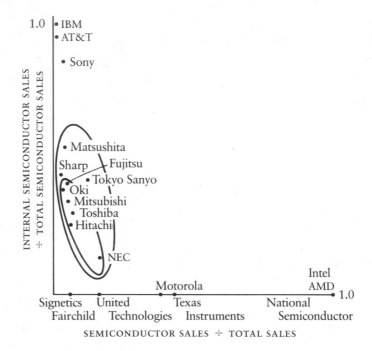

Fig. 6. Relation between internal semiconductor sales, semiconductor sales, and total sales of selected U.S. and Japanese companies, 1978-1980. For the Japanese companies the ratios of internal consumption of integrated circuits to total integrated circuit sales are taken from Masuda and Steinmueller's table 1. They are based on pre-1980 data. The ratios of total semiconductor production to total sales are from fiscal 1980; the values of semiconductor production for those companies were taken from the *Japan Economic Journal*, Dec. 23, 1980, p. 9; the values of total company sales come from company reports. The only exception is Matsushita for which only 1978 data were available. For the U.S.-based companies, 1980 total sales and semiconductor sales come from annual reports. We make several inferences to facilitate comparison.

United Technologies does not identify a semiconductor line of business. Semiconductor sales are included in the segment "Industrial Products and Services for the automotive, appliance, and other industries." The figure for this segment is reported here as semiconductors. On November 1, 1979, United Technologies acquired Mostek. In 1978 Mostek sales were $134,013,000—12.5 percent of United Technologies' sales attributed to Industrial Products and Services for the automotive, appliance, and other industries ($1,069,792,000). The fraction of semiconductor sales to total sales in the table is probably an overestimate. There are no figures available on internal sales of semiconductor devices within United Technologies, and we have no knowledge of major operations within United Technologies that buy principally from the operation. Consequently, we assume that internal semiconductor sales are negligible. Internal sales were, to our knowledge, negligible.

Signetics was acquired by Philips Corporation in 1975. To our knowledge there are no significant integrated circuit sales from Signetics within Philips. Internal sales were, to our knowledge, negligible.

Fairchild was acquired by Schlumberger in 1979. In 1978 Fairchild's total sales were $533,832,000, of which $383 million were attributable to semiconductors.

IBM did not sell semiconductors to other companies, and we have no estimate of the magnitude of semiconductor production. We do, however, know that IBM used all the semiconductors it produced.

AT&T, like IBM, uses all the semiconductors it makes.

Motorola makes no note of its internal sales of semiconductor products so we assume they are negligible.

National Semiconductor (NS) makes no note of its internal sales of semiconductor products so we assume they are negligible.

greater specialization in integrated circuits of the U.S. companies. It should be added that Philips (which acquired Signetics), Schlumberger (which acquired Fairchild), and United Technologies (which acquired Mostek) were all companies for whom integrated circuit revenues were not a large part of overall revenues.

Figure 6 shows the relative positions of the U.S. and Japanese companies according to these two measures. On these dimensions there seem to have been greater differences between the U.S.- and Japan-based companies as groups in 1980 than there were within the groups. Just as important, Figure 6 supports the notion that there is considerable variation among the companies in both industries. This variation and the competition among companies based in both countries was, we believe, an important feature of the industries.

Several natural groups of companies are suggested by the figure and by nature of the businesses.

1. Heavily diversified and somewhat vertically integrated electronics firms prominent in the state-of-the-art, high-volume memory market: Fujitsu, Hitachi, Mitsubishi, NEC, Oki, and Toshiba.

2. Heavily diversified and somewhat vertically integrated electronics firms not prominent in the state-of-the-art, high-volume memory market: Matsushita, Sharp, Sony, Tokyo Sanyo.

3. Moderately diversified and not vertically integrated electronics firms: Motorola, Texas Instruments.

4. Specialized integrated circuit firms: AMD, Intel, National Semiconductor.

5. Conglomerates (heavily diversified and not vertically integrated) selling integrated circuits: Philips (Signetics), Schlumberger (Fairchild), United Technologies (Mostek).

6. Diversified and vertically integrated producers of semiconductors: AT&T, IBM.

The Japanese firms fit the first two categories of "heavily diversified and somewhat vertically integrated" electronic firms.* Matsushita Electric consumed somewhat more of the semiconductors it produced

---

* Official exchange rates and inflation rates between and in the U.S. and Japan fluctuated significantly during the 1970's. The official indexes, of course, cover considerably more than semiconductor production. The price per logic element of integrated circuits has declined considerably over the period so the inflation indexes even have the wrong direction. Capital equipment prices are probably somewhat better represented by the economywide price series, and we do provide inflation-adjusted totals of capital expenditures in Tables 13 and 14. But even these are weak. We use statistics that are not adjusted for inflation unless the adjustment is explicitly mentioned. The exchange rate used throughout the chapter is 210 ¥ for $1 U.S. This, too, is an expedient compromise.

than did the main group of firms—enough more to make including it in the group not entirely satisfactory. Sony uses so many of its own semiconductor devices that it might well have been classified as a "diversified and vertically integrated producer" like IBM and AT&T. Sony was, of course, considerably smaller than those two in terms of assets and sales; all three had not diversified far from their main final product focus. In Figure 6 the loop enclosing all the Japanese firms' statistics does not include any U.S. firm's statistic. Note also the smaller loop that encloses all the major Japanese participants in the state-of-the-art, high-volume memory market. The major Japan-based memory participants consumed a smaller portion of their semiconductor production than did the Japan-based companies that do not participate.*

There was also a growing number of U.S.-based electronics firms that had established captive integrated circuit fabrication facilities in the early 1980's. Hewlett-Packard and Digital Equipment were among them. These companies developed integrated circuits primarily for some of their own needs, though in the early 1980's they still bought their requirements of the high-volume integrated circuits from outside vendors. Similarly, Japanese companies like Seiko (watchmaker) and Nippon Denso (automotive parts producer) began in the early 1980's to produce small volumes of special-purpose circuits. The size and frequency of such special-purpose captive production might have been expected to increase (see Chapters 2 and 3). However, because it would probably not have a major impact during the mid-1980's on the high-volume industry segment studied here, we have chosen to ignore them.

Another feature of the semiconductor manufacturers relevant to finance was the amount of non-semiconductor–generated retained earnings available for investment in the semiconductor operations. These funds, if available, might have been less expensive than external funds in both the U.S. and Japan because transactions costs of capital acquisitions were probably lower for internal than for external capital. The size of the non-semiconductor part of a company is not a completely reliable indicator of the magnitude of non-semiconductor internal capital available for semiconductor investment. That varies considerably depending on the nature of those other operations. Nevertheless, it is instructive to compare the U.S. and Japanese semiconductor firms along these dimensions.

Figure 7 relates total company sales (in billions of U.S. dollars) in

* We should note that Toshiba, Mitsubishi, and NEC are the only three of these Japan-based companies that are closely associated with keiretsu. We have not discerned significant competitive impact of these connections per se.

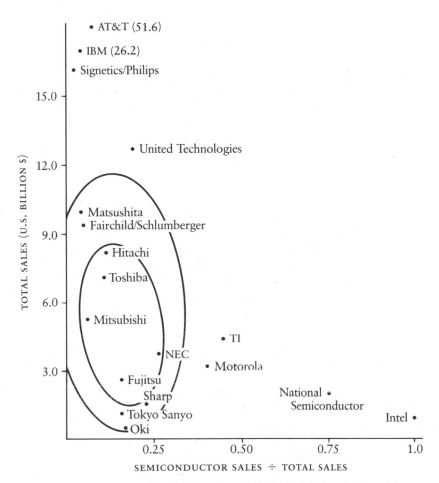

Fig. 7. Relation between semiconductor sales and total sales of selected U.S. and Japanese companies, 1980. Data for the value of semiconductor production for Japan-based companies for 1980 come from the *Japan Economic Journal*, Dec. 23, 1980, p. 9. (The exchange rate used in computations is 210 ¥ for $1 U.S.) Data for Sony were unavailable. Data for 1980 total sales of the Japan-based companies are private data supplied by Nomura Research Institute. Data for the total sales and semiconductor sales of the U.S.-based companies come from annual reports for 1980. We assume that the semiconductor sales of IBM and AT&T were zero. Estimates of semiconductor 1980 sales of Signetics and Fairchild were unavailable because they are now owned by non-U.S. based companies. We use Fairchild's 1.44 times total semiconductor sales in 1978 as an estimate of Fairchild's 1980 semiconductor sales. Schlumberger's 1980 total sales were translated at an exchange rate of 4.5 French francs per U.S. dollar. Signetics' 1980 semiconductor sales were taken as 1.44 times the Dataquest estimate of Signetics' 1978 sales; Philips' 1980 total sales were translated at an exchange rate of 2.13 Dutch guilders per U.S. dollar.

1980 to the percent of 1980 total sales (or value) in semiconductors. Roughly the same groupings of companies emerge as in Figure 6.

One smooth loop encloses all the Japan-based companies and only Fairchild/Schlumberger of the U.S.-based companies. In general, the value of each Japanese company's semiconductor production was a smaller proportion of company sales than for the U.S. firms that emphasize semiconductors. The sales in the semiconductor area were similar to those of the Japanese companies, but the U.S.-based merchant semiconductor businesses lay in companies with total sales smaller than those of many Japanese companies.

The additional implication here is that IBM, AT&T, and the conglomerates producing semiconductors had access to much larger internal pools of funds than any of the other companies—including all the Japanese companies.* The specialized U.S. semiconductor firms had access to much smaller internal pools of funds than the leading Japanese companies against whom they competed most closely in the high-volume market segments. These gross differences in size of the internal capital pool to which companies had access corresponded to the difference in degrees of severity of capital cost, and availability for future growth was discussed as a problem more by the U.S. managers than by the Japanese managers we interviewed. Among the U.S. firms the specialized integrated circuit manufacturers seem to have been most vocal.

The generalizations we make here are of course "snapshots" of an industry in transition. Change has been a fundamental characteristic of the industry, and international competition increased the rate of change. We use this snapshot as a guide in the different effects financing might have had on the firms' competitiveness.

Global Competition

The companies we have been discussing—with their different resources—were the principal actors in the high-volume digital integrated circuit market during the late 1970's and early 1980's. Throughout the short history of their industry, they had as individual companies held long-term views of themselves and their industry. They had been

* We should note that NTT, which did quite a lot of the basic research and development in semiconductors for telecommunications in Japan, bought its semiconductors from outside vendors that are in our sample. NTT announced in 1982 the formation of a subsidiary Nippon Electronic Engineering Co., to "design, prototype, and test custom large-scale integrated circuits" eventually for outside customers (*Electronics*, June 2, 1982, p. 83). NTT might eventually be more comparable to IBM and AT&T than to the current Japan-based companies.

willing to price low in order to gain market share and experience to use in future products. They had been willing to invest a portion of their revenues in R&D larger than that of most companies in either economy. As their products defined new industries and their industry sales mushroomed, the cost of their products declined. The properties of the global competition emerging in semiconductors during the late 1970's and the 1980's would result then, from the combination of the competitive strategies of the participants, the role of government, and the continuing rapid rate of advance in semiconductor technology.

Changing capital requirements. In this section we discuss some evidence that the capital necessary to participate in this industry was increasing during the late 1970's and early 1980's, and might continue that trend.

Chapters 2 and 3 discussed the nature of capital equipment needed for state-of-the-art integrated circuit processing. The costs of that equipment increase substantially as technology advances. Here we discuss historical trends in sales and investment. Together, we argue, these two bodies of evidence make a strong case that (1) entry requirements to the high-volume digital integrated circuit industry segments, at least, were increasing and would continue to increase, and (2) the average capital investments per sales dollar also were increasing and would continue to increase.

The most striking trend was, of course, the growth in capital expenditures. Tables 16 and 17 present estimated capital spending on semiconductor operations for eight important merchant manufacturers in the U.S. and nine important merchant firms in Japan. Between 1973 and 1976 nominal capital expenditures of the U.S. industry increased only slightly and the average growth was slowed because of the recession of 1974-75. Although the recession also had its effects on capital spending in Japan, nominal capital spending for that industry doubled between 1973 and 1976. Of course, in 1976 the U.S. industry spent about $120 million more than the Japanese industry. Between 1976 and 1980 total capital spending by the U.S. companies almost quadrupled. During the same period total capital spending by the Japanese companies almost tripled. During the entire period the Japanese companies went from spending about one-third the U.S. companies' capital expenditures in 1973 to about one-half in 1980. Overall, the combined capital expenditures of the Japanese companies grew more rapidly than those of the U.S.-based companies. According to the U.S.-based Semiconductor Industry Association, this disparity had increased by 1982.

TABLE 16

Estimated Capital Expenditures of U.S.-Based Merchant Manufacturers
(Semiconductors Only), 1973-1979

(Millions of dollars)

Company	1973	1974	1975	1976	1977	1978	1979	1980
Advanced Micro Devices	6	5	1	5	7	22	42	65
Fairchild	35	41	20	36	15	23	58	95
Intel	9	13	11	32	33	85	82	125
Mostek	6	10	3	10	24	19	42	85
Motorola	41	71	21	33	43	72	159	175
National Semiconductor	21	20	17	26	31	49	70	120
Signetics	22	20	4	10	18	40	50	80
Texas Instruments	64	68	36	62	88	115	190	220
Other U.S. companies	137	162	81	144	169	255	372	415
Total spending	341	410	194	358	428	680	1,065	1,380

SOURCE: Annual Reports DATAQUEST, Inc., October 1978, March 1980, and December 1980.

TABLE 17

Estimated Capital Expenditures of Japan-Based Manufacturers
(Semiconductors Only), 1973-1980

(Millions of dollars)

Company	1973	1974	1975	1976	1977	1978	1979	1980
Fujitsu	24.8	10.5	3.8	6.7	15.3	50.6	69.7	117.4
Hitachi	25.8	27.7	6.7	30.5	21.9	42.9	64.9	100.2
Matsushita	NA	NA	4.8	15.3	23.9	21.9	47.7	86.9
Mitsubishi	15.3	19.1	17.2	23.9	25.8	25.8	34.4	42.9
NEC	15.3	25.8	15.3	50.5	39.1	66.8	117.4	138.4
Oki	3.8	3.8	4.8	14.3	18.1	14.3	23.9	56.3
Sharp	2.9	4.8	4.8	17.2	4.8	8.6	38.2	37.2
Tokyo Sanyo	1.9	6.7	1.9	10.5	8.6	6.7	19.1	34.4
Toshiba	21.9	8.6	13.4	61.1	17.2	25.8	42.9	56.3
Total spending	111.7	107.0	72.7	230.0	174.7	263.4	458.2	670.0
Total adj. by WPI (1975 = 100)	150.8	109.8	72.5	223.4	163.2	195.7	409.5	508.8

SOURCE: The raw statistics here come from the *Japan Economic Journal*; the WPI (wholesale price index) come from the Bank of Japan.

We should note that we focus on 1980 as a benchmark because it was a year of "good business." In 1981 a recession hit the industry; sales and investment levels were both lower than the trend. Our interest here is how the availability and cost of finance to different companies might affect their competitive viability. This is more likely to be a problem if capital requirements are high. The kinds of financial forces that could affect competition will be most apparent if the projections are on the high side. So, with the caveat that our forecasts of capital requirements are intentionally high, we proceed using 1980 and the trends of the 1970's to guide our estimates.

The generalization that the Japanese industry's capital expenditures grew more rapidly though they remained below the U.S. level in 1980 is not uniformly true at company level. The top four U.S. spenders on capital in 1980 were (in descending order) Texas Instruments, Motorola, Intel, and National Semiconductor. Their capital expenditures in 1980 ranged between 3.5 and 5.3 times their expenditures in 1976. In 1976 Fairchild was the second largest U.S. spender on capital, and in 1980 it was a distant fifth.

In Japan the four largest spenders on capital for their semiconductor operations in 1980 were (in descending order) NEC, Fujitsu, Hitachi, and Matsushita. By 1980 their capital expenditures had increased to between 2.7 and 17.6 times their levels. Fujitsu's expenditures grew most rapidly; its 1980 capital expenditures (converted at 220 ¥ = $1 U.S.) were the second highest in Japan and greater than all but three U.S. companies. (Depending on the exchange rate applied, they approximately equaled Intel's 1980 capital expenditures.)

Not all the Japan-based companies' capital expenditures grew more rapidly than those of their U.S. competitors. In fact, Hitachi and NEC both had ratios of capital expenditures in 1980 to capital expenditures in 1976 lower than those of the four largest U.S. spenders. The pattern of greater capital expenditures increases in Japan seems to reflect the big increases in investment of Fujitsu and Matsushita. Fujitsu, of course, was a big and early participant in the 64K RAM market. Of the big four capital spenders in both countries only Matsushita seems not to be a strong potential contender in that market.

Another measure of growth in capital expenditures also cuts across the two countries' industries rather than between them. The magnitude of the increase in capital expenditures in 1980 over 1976 was highest for National Semiconductor and Fujitsu. Figure 8 shows the levels of semiconductor-related capital expenditures and the corresponding annual increase in capital expenditures for 1980. Texas In-

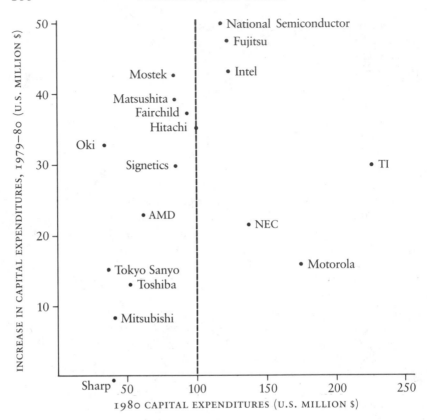

Fig. 8. Level and growth of capital expenditures of selected U.S. and Japanese companies, 1979-1980. (Source: data from Tables 16 and 17. Yen were converted at 210:1.)

struments and Motorola stand out as the largest spenders with NEC as number three. National Semiconductor and Fujitsu increased their expenditures most rapidly in 1980.

Just as the industry-level data may be too highly aggregated to provide a picture of the competitive strengths, the company-level data may also be. Certainly, the companies did not all compete on all the same product markets. But a number of these companies seemed to be competing against one another in what they said were critical segments of the large semiconductor market. One of the most notorious of those segments is the emerging 64K random access memory (RAM) market. According to *Electronics*, all the companies with capital expenditures greater than or equal to Hitachi's were either in or close to production of that product in September 1981.[13]

The other three companies moving close to production at that time were Oki, Mitsubishi, and Toshiba. These three companies had significantly lower levels of capital expenditures than several U.S.-based companies that were not yet in the market. Mostek and Fairchild were also preparing 64K RAMs. Mitsubishi and Toshiba may have concentrated their digital semiconductor resources in this product market, unlike the other U.S.- and Japan-based companies that had broader product lines. (Toshiba had concentrated on consumer-oriented circuits.)

Figure 8 seems to support the notions that the 64K RAM market was indicative of the larger state of competition among U.S.- and Japan-based companies and that the capital expenditures of those companies were related to their strengths in that marketplace. The dotted line in the figure separates the strong competitors in the 64K RAM market from the others.* These also happened to be the companies with the largest levels of capital expenditures.†

This suggests that the three leading U.S. companies in terms of sales and capital expenditures and the three leading Japanese companies in both those terms were competing closely during the early 1980's. It seems clear that Fujitsu, in particular, was investing at a very rapid pace as compared with its chief American competitors. However, the notion that the Japanese companies were in 1980 expanding their capital expenditures much more rapidly than their U.S. competitors seems to mean mainly that the largest spenders in both countries— NEC, Motorola, and Texas Instruments—were not accelerating their spending as fast as the smaller companies.

If one assumes further that the top seven companies continued to increase their capital expenditures for the next few years at the rates at which they increased them between 1979 and 1980, the result is that TI and Motorola drop relative to the more rapidly increasing spenders like Fujitsu, Hitachi, Intel, and National. Of course, whether and how this would occur was unclear in the early 1980's. In particular, Intel and National were concerned about their continuing to have access to increasing amounts of capital.

Another related piece of the puzzle is the capital intensity of sales. In 1973 the capital expenditure to sales ratio of the U.S. firms was 12

---

* Oki was preparing to introduce a 64K RAM product, and its financial characteristics were of course different from those of the other Japanese companies. Oki had dangerously poor performance. It had been mainly a supplier of telecommunications equipment to NTT, and NTT seemed to be in the process of revitalizing the company.

† Fairchild's capital expenditures have only been estimated since Schlumberger does not report them. It is certainly possible that Fairchild's actual capital expenditures in 1980 would put the company to the right of the dotted line.

percent. By 1979 this ratio had climbed to 15 percent, and by 1980 it reached 16 percent. For the nine main Japanese firms the ratio of combined capital expenditures to combined value of semiconductor production was 16 percent in 1979 and 17 percent in 1980. Most industry experts agree that this reflects an increase in the capital required to produce an advanced technology sales dollar. In 1980 the independent merchant semiconductor companies based in the U.S. had asset to sales ratios of between 0.20 (National) and 0.38 (Intel). If the requirement for increased capital per sales dollar was correct, then these companies would require considerable increases in investment to move these ratios to 1.0, for example, or better for advanced products.

The Japan-based companies did not report their semiconductor-related fixed assets, and they were, of course, diversified into other businesses (some of which were capital-intensive). So it is impossible to be sure of their 1980 fixed asset to sales ratios. Judging from the change in new capital expenditures, however, the capital intensity of sales seemed to be increasing generally and increasing more rapidly in Japan than in the U.S. The current difference could be explained at the company level by the proposition that a larger portion of Japanese semiconductor production was done by the companies like Hitachi and Fujitsu whose semiconductor businesses were growing rapidly as opposed to the companies with historically larger capital expenditures like Motorola and TI.

Company-level data do seem to support this explanation. Figure 9 relates 1980 semiconductor sales, or the value of semiconductor production, to semiconductor capital expenditures. As semiconductor sales, or production value, rose the ratio of capital expenditures to sales also rose. The dotted line separates points of more rapid rise— several Japanese companies—from those of lower rise—the American firms and Toshiba. What this means for relative competitiveness is not clear. But it does suggest that, *for their size* (or competitive position), several of the merchant producers based in the U.S. were not increasing their investments as rapidly as were their Japanese competitors. Except for Texas Instruments and Fujitsu the magnitude of this difference was probably not significant. Fujitsu, in particular, stood out as a company investing a great deal relative to the value of its production. Fujitsu and Matsushita might have been investing in entering the semiconductor market more than the other Japanese or U.S.-based companies.

The above statistical comparisons all suggest that capital expenditures were increasing for all the semiconductor manufacturers and

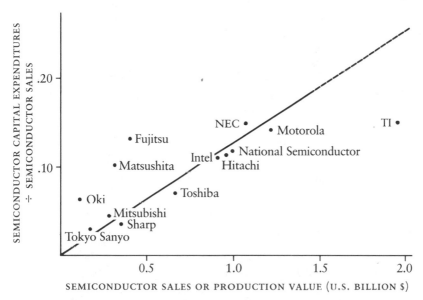

Fig. 9. Semiconductor sales and capital expenditures of selected U.S. and Japanese companies in 1980. (Sources: for the U.S.-based companies, annual reports; for the Japanese-based companies, *Japan Economic Journal*.)

that they were increasing relative to sales for all the successful competitors in high-volume product segments. Capital expenditures did not seem to be increasing much more rapidly in relation to sales for most Japanese companies than for most U.S. companies. Rather, Fujitsu seemed to be increasing its capital expenditures more rapidly than all its competitors, especially in relation to the valuation of its production. Motorola, TI, and NEC were increasing their annual capital expenditures less rapidly than their competitors, but they had already established a much higher level of expenditures.

All the companies in the industry spent more on capital each year. The seven leading companies were spending at least $100 million a year in 1980 and increasing that by roughly $20 million to $50 million a year. A pressing question for the industry was how much capital requirements would increase over the next five or ten years.

Industry experts explain that the capital equipment required to produce a functional element on an integrated circuit had increased in cost and would continue to increase while the selling price of the functional element had decreased and would continue to decrease. This trend also meant that the cost of establishing a new plant for fabricating state-of-the-art integrated circuits had increased and would continue to do so. The Signetics MOS plant in Albuquerque, New Mexico, was budgeted in 1981 at $100 million.[14] At about the same time, Motorola reportedly spent $64 million for its 64K RAM plant in Austin, Texas, and Oki Electric reportedly spent about $60 million for its new 64K RAM plant in Miyazaki, Japan. We heard of no plant planned during the early 1980's for fabrication of LSI ICs in commercial quantities in either country that cost less than $50 million. At that time, Fairchild was reputedly preparing to invest in a bipolar plant on Kyushu with the expectation that the total investment there would be $100 million. The plants were also to a much greater extent than previously located in countries other than that in which the company was based.

If these technical projections held, the capital intensity of production would increase considerably from the levels of the early 1980's. Trend extrapolation can provide some rough notions of what these companies' capital requirements might be during the next five to ten years. One such extrapolation can be based on Dataquest estimates. They suggested in the early 1980's that by 1990 Japan's share of the world market for semiconductors would increase from 40 percent to 45 percent. Given their total market projections, this meant that the U.S. merchant industry would produce $37.2 billion of semiconductors while the Japanese would produce $17.1 billion. If capital intensity remained at about 17 percent, the U.S. firms would spend $6.3 billion and the Japanese firms $2.9 billion on capital; if it increased to 20 percent, the U.S. firms would spend $7.4 billion and the Japanese firms $3.4 billion. In contrast, in 1980 the U.S. companies spent $1.38 billion and the Japanese companies $792 million.

By 1980 semiconductor sales worldwide had grown at about 20 percent per year for over twenty years. There had been recessions and several major technological changes, but the trend held on the macro level. For our purposes of estimating likely capital requirements for companies in the industry, we shall assume that industry sales would continue to grow at 20 percent annually. We shall take into account the comparatively rapid increases in capital expenditures undertaken by Fujitsu, Intel, and National Semiconductor.

The challenge of raising capital might have been complicated by precise timing requirements for the capital. There is statistical evidence that introducing a new semiconductor product earlier than competitors, all else being equal, can have a lasting positive effect on market share and profits.[15] This could be because of learning in production, the possibility of creating loyal customers, and/or prestige. Thus competitive potential might depend on the timing of capital acquisition.

Many U.S.-based semiconductor manufacturers may have been facing the challenge of timing during the early 1980's. Their announced capital expenditures for 1981 and 1982 followed the increasing trend suggested above. But their realized capital expenditures were somewhat less than planned. Intel and Signetics delayed openings of new U.S. plants in 1981, and National Semiconductor postponed (or abandoned) its plans to build a new plant in Texas. Dataquest was quoted in early 1982 as saying that capital expenditures of U.S. merchant semiconductor companies would probably decline 6 percent in 1982 from a total of $1.41 billion.

This decline may have been intentional during the 1982 recession. It appears that total semiconductor-related capital expenditures worldwide rose in 1982 by only 34 percent in contrast to 50 percent the previous year. So Japanese firms, like U.S. firms, may have been holding down new investments. R&D expenditures are counted for tax purposes but only yield returns some time later; they were not reported uniformly by line of business in official company reports. Believable estimates of R&D applied to semiconductors are difficult to obtain. According to their annual reports, U.S.-based companies between 1972 and 1980 spent between 6 percent and 12 percent of their revenues on R&D. In the period 1977-80, these companies spent an average of 8 percent of their sales on R&D. The proportion of their *semiconductor* revenues that they spend on *semiconductor* R&D may of course differ considerably from these figures.

This percentage is high compared with the 3 percent average for all U.S. firms, but Japan-based semiconductor manufacturers were reported to have spent 16.9 percent of their revenues from integrated circuit sales on R&D between 1973 and 1978. In 1979 this would have been $524 million (210 ¥ = $1 U.S.) on semiconductor-related R&D. In 1980 this would have amounted to $742 million.

Cost and availability of capital effects. In the context of severe price competition from forward-looking firms, differences in the cost of capital to firms based in the U.S. and Japan could have a significant effect on the outcomes of the competition.

The most straightforward way in which this effect would occur would be for the cost of semiconductors to rise in the country with higher capital cost. This would in turn raise the break-even price of the companies based in that country, and it would create a situation similar to that observed in parts of the semiconductor industry during the early 1980's where costs of production were lower in Japan than in the U.S. If product markets were price competitive—as markets have been—then the companies with higher capital costs would be earning a lower return on their investments than they otherwise would. On products for which other costs of the companies were identical, the companies with higher capital costs would lose money.

If the strategies of the semiconductor companies were simply to make profits on each product, then the companies with higher capital costs would not enter product markets in which they had cost structures and/or introduction dates similar to their lower cost of capital competitors. They would withdraw from product markets as quickly as possible once they began losing money.

But the competition in the high-volume digital integrated circuit segment of the semiconductor industry required participation in an ordered sequence of related products. That is, technical experience in processing simpler design products (like RAMs) in a new generation of products would be necessary in order to design and process more complicated products in that generation like microprocessors. And experience in one generation would be essential to effective technical performance in the next.

This meant that once a company left the industry the costs of reentering would be very high. So, anticipating that they would make profits on a later generation, companies might remain in unprofitable product markets. The present value of the losses from continuing operations might be less than the present value of reentry costs after exit.

This dynamic process sets up the possibility of a protracted competition during which the companies with high capital costs, all else being equal, lose money. Of course, those companies could still make profits during the competition, but they would do so only by being earlier, or better, than their competitors with low capital costs.

During the early 1980's it seemed possible that the U.S. semiconductor companies were facing the prospect of being the high capital cost companies in just such a competitive market with excellent technical rivals in the Japanese companies. As long as the Japanese and U.S. financial systems were separate, this possibility would remain a competitive disadvantage in the years ahead.

Availability and sustainability of capital effects. In the context of a prolonged series of competitive encounters—or races—in semiconductor markets, the availability of capital could have just as significant an impact on competition as its cost. Suppose that capital costs adjusted for risk and technology and the assets were identical for the U.S. and

Japan competitors. The race would be won by the companies that stayed in the longest. The prize would go to the investors in the winners.

The investors in all the competitors would increase the "price" at which they would invest as they perceived the risk of losing to increase. In particular, the individual investor's perceived probability of losing would increase with the probability that the company's other investors would withdraw support of its participation in the race. During a race this probability was likely to be higher for U.S. investors investing from outside in U.S. semiconductor firms than for Japanese semiconductor firms. This followed because the Japanese city banks were less likely than U.S. investors to be the last to realize that the company's other investors had withdrawn support; and this, in turn, followed because MITI and the city banks monitored *all* the investors in the semiconductor manufacturers: similar quality of monitoring of a merchant U.S. firm's many investors did not occur.

So for Japanese semiconductor firms as a group there was a greater possibility of maintaining financing during a prolonged "race" than for U.S. firms, all else equal. In an industry where staying in business would be much less expensive than reentry, this aspect of the Japanese financial system could provide a critical competitive edge to Japanese firms.

Of course, there is another source of greater long-term accessibility to capital that might cut across the two semiconductor industries— that is, access to capital from sources internal to the company but external to the semiconductor operations. Companies like Schlumberger, United Technologies, and Philips acquired merchant U.S. semiconductor companies during the late 1970's and early 1980's. Presumably they did this less for the profit prize at the end of the race than for the security of their sources of integrated circuits and for the high profits they expected from technological spillovers in their related businesses. These types of companies might supply internal capital during the race to their semiconductor operations. They might even do so for a longer period than the Japanese financial system would.

Venture capital effects. Capital for new ventures was, of course, much more easily obtainable in the U.S. during the late seventies and early eighties than in Japan. Suppose we were wrong in anticipating a highly contested race in the high-volume digital integrated circuit markets, and suppose that instead there ensued considerable technological ferment that lowered the capital requirements for entry to the industry. Then new entrants in the U.S. might well have the competitive advantage over the big Japanese and U.S. companies. Without a source of venture capital in Japan there would be no comparable new entrants in Japan.

Governments' effects. Direct action by either government in providing subsidies to semiconductor manufacturers could, of course, alter the cost and availability of capital to their firms during a race. Such subsidies, if undertaken by one government and not the other, could provide a significant advantage to the recipient companies.

Probably the biggest effect of government action was indirect. Governments played a fundamental role in constructing and maintaining the institutional infrastructures that form the basis of the important financial differences between the United States and Japan. Governments also pursue macroeconomic policies that, as we saw above, can have a considerable impact on the cost of capital and the amount of capital available to corporations.

Conclusion. It is important when considering our discussion of the potential competitive effects of differences in the financial systems of the United States and Japan to keep in mind our ubiquitous *ceteris paribus*. Technology development, production, and marketing operations capabilities were "held constant" in various ways convenient to make our arguments clear. In the semiconductor industry, of course, these resources are not so easily and conveniently "held constant." In particular, in the sort of "race" over products we envision here each company's real resources and management affect its future competitive resources.

Finance enters the competitive picture when access to capital is not uniform for the competitors in an industry. The evidence discussed above suggests that it was not uniform for the competitors in the global semiconductor industry during the late 1970's and early 1980's. Moreover, the effects of financial differences on competitive outcomes are even greater for industries like the semiconductor industry because entry becomes more difficult and expensive over time.

The effects of the financial differences we discuss in this chapter on the semiconductor industry were potentially large. Although the Japanese financial system was becoming more open, it was doing so only very slowly. The movement was so slow that most observers would not have expected the opening to attenuate the competitive effects of financial differences during the 1980's.

SIX

Conclusions

Daniel I. Okimoto

Each chapter in this book has dealt with a different but important aspect of the U.S.-Japanese competition in semiconductors. The topics may seem diverse; but there is a common thread running through the chapters and tying them together. It is the shared view that all aspects of the bilateral competition—technology, government and politics, and finance—must be viewed as interdependent parts of national systems. The competition is not simply between American and Japanese corporations: IBM and Fujitsu operate within political-economic systems that greatly affect their international competitiveness. To assess the relative strengths and weaknesses of American and Japanese sides, therefore, requires that the various components of the semiconductor competition be fitted together in a broad scheme.

Figure 10 shows the relationship among the various factors at work in the broad areas of technological innovation, finances, and government policies.[1] To a large extent, company and sectoral performance depend on what happens in these three domains, and the United States and Japan possess strengths and weaknesses in each.

Technological Innovation

Technological innovation has been one of the hallmarks of the semiconductor industry. The speed, scope, and complexity of technological change set semiconductors apart from "smokestack" industries. No physical limit is yet in sight.[2] Opportunities for continuing progress, be it in the form of breakthroughs in product design and functions, or incremental advances in production technology, will remain open for commercial exploitation. For companies that successfully bring products to market first, the chances of reaping big profits, or of establishing market dominance, serve as powerful incentives.[3] If second-to-

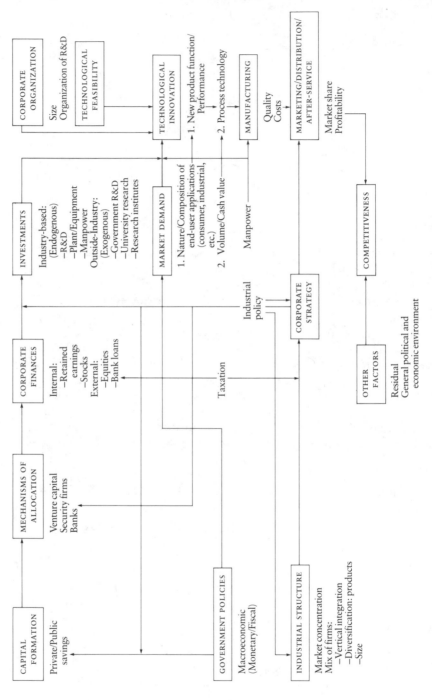

Fig. 10. Determinants of semiconductor industry competitiveness.

market firms improve even marginally on new products, or find ways of lowering costs through better production technology, they can also catch the swelling tide of market demand.[4]

If innovation is the sine qua non of commercial competitiveness, what determines it? The Japanese are renowned for ingenuity in production and process technology and product improvements, but they are said to lack originality. What factors seem likely to affect their capacity to innovate? This study suggests that two factors are especially critical: market demand and rates of investment.

Market demand. Chapters 2-4 have shown that market demand stimulates innovation in two ways: (1) through the final products (end-uses) into which semiconductor devices are installed, and (2) through the sheer volume and cash value of semiconductor production. The authors of these chapters agree that differences in the nature and pace of technological innovation in the U.S. and Japan are attributable, in part, to differences in the mix of products for which semiconductor devices have been used (see Fig. 11). In the U.S., the largest proportion of market demand has come from computer and industrial sectors; and in the early phases of development, the military generated a significant portion of demand. America's mix of end-user demand has pulled semiconductor technology in the direction of ever more sophisticated and densely packed digital chips; semiconductor components have increasingly taken on the features of whole systems.

In Japan, the composition of market demand has reflected early decisions to enter the semiconductor arena via consumer electronics, especially hand-held calculators.[5] Consumer goods account for the largest share of semiconductor production. Computer and industrial applications are rising but still nowhere near U.S. levels. Japanese companies produce a higher proportion of linear ICs, estimated to be 26 percent in 1980, for use primarily in consumer goods.[6] (See Table 18.)

Linear ICs have not been at the forefront of technological (mainly MOS digital) advances, nor are they considered a good springboard for achieving technological advantage in the world IC market. As Table 18 shows, in both Japan and the United States bipolar linear IC production has been declining relative to total production since 1978, and in absolute terms also, linear IC growth rates are lowest of the four categories. Though Japanese production still has a way to go before dropping to U.S. levels, the downward trend is clear, particularly in comparison with the rapid rise of MOS memory and logic devices. In three years, MOS memory production in Japan rose from 18 percent, only about half that of linear ICs, to 27 percent, or slightly higher; if

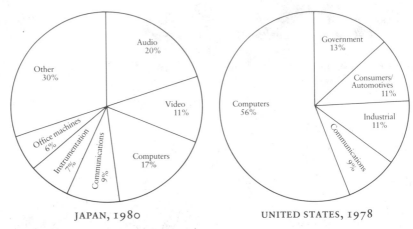

JAPAN, 1980   UNITED STATES, 1978

Fig. 11. Semiconductor end-user applications in Japan (1980) and the U.S. (1978). (Sources: for Japan, MITI; for the U.S., Nico Hazewindus, *The U.S. Microelectronics Industry* (New York: Pergamon Press, 1982, p. 82).)

the trend continues, Japanese IC production will be moving into more sophisticated, "cutting edge" product areas, reflecting shifts in the composition of original equipment manufacturer (OEM) demand, both domestic and international.

If Japan's computer industry continues to grow at its present pace, computer and industrial demand should eventually overtake the demand for consumer goods. In that event, one can expect that the same market forces that promoted new inventions in U.S. semiconductor technology will also be at work in Japan: that is, Japanese firms will have to innovate more vigorously, especially in new product design and functions, in order to stay competitive.*

Less dramatic than new product design but no less important commercially is process technology. Process improvements tend to be incremental in nature, with no single advance being very significant by itself; but the cumulative value is often great. It can spell the difference between success and failure in the marketplace. Process technology is the means by which second-to-market firms secure an expanding market position. Japanese companies have been able to advance rapidly from positions far behind largely because of their concentration on,

* Of course, one must be careful not to overstate the relationship between innovation and market demand. The relationship becomes tenuous when connections begin to be made between potential demand and innovation; the number of products for which potential markets exist is virtually unlimited. See David C. Mowrey and Nathan Rosenberg, "The Influence of Market Demand upon Innovation," in Nathan Rosenberg, *Inside the Black Box* (New York: Cambridge University Press, 1982), pp. 193–241.

TABLE 18

U.S. and Japanese Integrated Circuit Production, 1978-1981

Type of circuit	Percent of total production			Absolute growth rate
	1978	1979	1980	1978-81
United States				
MOS memory	24%	28%	31%	27%
MOS logic	25	25	23	22
Bipolar digital	31	29	23	23
Bipolar linear	20	18	15	16
Japan				
MOS memory	18	25	27	49
MOS logic	34	32	33	44
Bipolar digital	16	15	14	28
Bipolar linear	32	28	26	27

SOURCE: Nihon Denshi Kikai Kogyokai, *IC Shuseki Kairo Guidebook, 1981* (Tokyo: Nihon Denshi Kikai Kogyokai, 1981), pp. 18, 21.

and successes in, incremental improvements in process technology. For many years, from the 1950's to the mid-1960's, the strict demands of semiconductor production bedeviled the Japanese industry. Only after sustained trial and error and the importation of U.S. technology were Japanese firms able to catch up; but by about 1970 the Japanese had come to possess world-class strength in manufacturing technology, the sine qua non of commercial competitiveness.[7]

Process technology is closely related to market demand in terms of volume and cash value. Improvements in process technology help to expand market demand, which in turn leads to greater efficiency as a consequence of learning by doing. Learning curve theory is thus a function of production volume. It is the reason semiconductor companies have been able to reduce costs by roughly 30 percent with every doubling of output. Lower costs can give companies a competitive edge in pricing, which can further increase output. Volume demand is thus capable of triggering a "virtuous cycle": reductions in per unit costs (economies of scale), efficiency improvements (learning curve), price competitiveness, growing market share, and reinvestment in R&D and new plant and equipment. If a company takes advantage of its opportunities, it can ride this "virtuous cycle," thereby putting pressure on its competitors and positioning itself to catch the next product cycle.

It is no accident that Japanese companies have excelled at manufacturing technology. They are willing to assign some of their best engineering talent to production, realizing that it is the key to increases in productivity, cost reduction, and price competitiveness. The emphasis is probably the result not only of the fierce competition between Japanese firms, with production efficiency one of the main determinants of market position, but also of high and sustained levels of investment in new plant and facilities and of the general manufacturing philosophy of a "catch-up" nation that has had to concentrate on productivity gains as the quickest way of overtaking foreign leaders. A comparative study of Japanese and American management has found that Japanese companies rank production factors at the top of strategic priorities, whereas American firms place the greatest emphasis on the development of new products.*

The stress on process technology makes sense, because it builds upon and exploits the particular strengths in Japanese corporate organization: lifetime employment, a spirit of teamwork, close communications, the high quality of assembly line workers, utilization of quality control techniques, capital availability, and relatively low union resistance to automation. Lester Thurow believes that because America has not paid enough attention to process innovation, U.S. industry has fallen behind Japan in the productivity race. He thinks America's system of R&D betrays a "substantial bias" toward the development of new products over production technology, and accordingly advocates greater government support for process R&D, in terms both of using public funds and of providing proper tax incentives.[8]

The military orientation of so much of federal R&D may explain the product R&D bias. Because the Defense Department's overriding concern is weapons performance, its R&D contracts tend to emphasize the design of new and better components and systems rather than process refinements. Of course such R&D underwriting often "spills over" into support for process refinement: the development of a denser RAM chip, for example, may require new process technology (e.g., implantation-through-metal, or epitaxial silicon single-crystal growth technology) in order to shrink the submicron widths separating circuits. If new products depend on breakthroughs in manufacturing technology, the military has a stake in solving production bottlenecks, and to the extent that advances in production technology lower the costs of military procurements, the Pentagon has an incentive to see that they are made. Nevertheless, the military's first priority is, and must be, the develop-

---

*Tadao Kagono, Ikujiro Nonaka, Kyonori Sakakibara, and Akihiro Okumura, "Japanese Inductionism vs. American Deductionism: A Comparative Analysis of Strategy, Structure, and Process," unpublished paper, March 1983, p. 3. For both sides, of course, product and production rank at or near the top; only the ordering is reversed.

ment of high-performance weapons hardware. If military R&D contracts lead to a lowering of production costs or to the commercial competitiveness of contract companies, it is a secondary (and largely unintended) consequence. The Pentagon is not in the business of trying to maximize civilian spillovers from military R&D; national defense is its responsibility, not industrial policy.

MITI's attitude is just the opposite. Its main responsibility is to facilitate efficiency and industrial competitiveness, not military security, and its R&D contracts are consciously aimed at promoting commercially optimal technology. Production technology falls squarely within the realm of legitimate government support. Furthermore, Tokyo seems less wedded than Washington to the proposition that process R&D ought to be left to the private sector. Although it is true that individual companies should have strong incentives to invest in production technology, since they themselves stand to gain from finding ways of reducing costs, the Japanese government believes that there are circumstances under which it ought to act to facilitate efforts in the private sector, particularly when Japanese firms clearly lag behind foreign pace-setters and suffer significant disadvantages in the international marketplace. To minimize the chances that only a small circle of firms will benefit disproportionately, MITI normally makes government-sponsored R&D patents widely available to all companies on an equitable basis. If manufacturing efficiency is enhanced thereby, MITI believes that Japanese industry, and ultimately the public at large, will be the benefactors.

Chapter 3 pointed out that Japan's VLSI project was oriented toward progress in production technology, and that perhaps its major achievement was electron-beam lithography. Phase III of America's VHSIC project is also oriented toward production technology, but it is earmarked to receive the smallest amount of money: $57 million out of a revised budget of around $300 million, or less than 20 percent of the total, spread out over the project's seven-year life-span.*

A noteworthy trend, which bears close observation over the next several years, is the rapid progress being made by Japanese equipment manufacturers, particularly in the areas of assembly and testing. Until around 1980, Japanese semiconductor producers had purchased the

---

* The VHSIC project ran into bottlenecks three years after its inception, when integrating semiconductor components into systems turned out to be more troublesome than expected, perhaps because of the project's explicit emphasis on component functions and performance. Cost overruns, a chronic problem in military R&D, raised the budget from an original estimate of around $200 million to a subsequent request for upward of a half a billion, and Phase I is still under way.

bulk of their equipment from U.S. manufacturers. But Nikon and Canon have since developed state-of-the-art equipment in such areas as photolithography and plasma etching equipment, and other manufacturers are making impressive strides in semiconductor manufacturing equipment.* The Japanese are mounting a determined effort here not only because it is a lucrative market but also because they do not want to be too dependent on American manufacturers; furthermore, they fear that some of the most advanced equipment, developed under Defense Department contracts (e.g., VHSIC), may not be available for export.[9]

In Japan, semiconductor equipment producers and systems firms have very close, sometimes overlapping, ties. NEC and Hitachi meet over half of their equipment needs through internal production. Moreover, NEC owns over 50 percent of Ando Electric; Hitachi owns Shinkawa and Kokusai Electric; and Fujitsu owns Takeda Riken. The Japanese are said to be behind in wafer fabrication equipment (e.g., ion implantation, metalization, diffusion, etc.), but such areas are not as central to components performance and yields as photolithography and testing, where the Japanese are already strong and are making rapid strides.[10] U.S. equipment firms are worried that the speed of Japan's progress has troublesome implications not only for the equipment market but for competitiveness in components production. The pervasiveness of vertical integration in Japan—from equipment manufacturing to components and systems production for a number of leading firms—also raises questions about potential commercial and production leverage. If, for example, Japanese corporations develop state-of-the-art equipment, will they sell such equipment to competitors? Or will they withhold them? Will experience gained in equipment manufacturing give them any extra edge in production know-how?

Under pressure from Japan's emphasis on process technology, some American companies have made adjustments in their approach to production—ranging from relatively easy organizational changes to more ambitious consolidation of ties with vendor networks. Hewlett-Packard, for example, has moved to solve interface problems between R&D and production by improving communications between R&D and production divisions not only in the early phases of design but also once the production phase has begun.[11] A few members of the

---

* The leaders are: *plasma etching*, Kokusai Electric, Tokuda Seisakusho; *testing*, Takeda Riken, Ando Electric, Kokusai Electric; *assembly*, Shinkawa Kaijo Denki; *metal deposition*, Tokuda Seisakusho, Anelva, Ulvac; *diffusion and deposition*, Kokusai Electric.

R&D team often move over to the production side in order to ensure that manufacturing proceeds smoothly; communications between R&D, production, and marketing, usually close in Japanese firms, have been improved at Hewlett-Packard and other U.S. companies. Hewlett-Packard is also trying to place relations with its large network of vendors on a stronger footing. Instead of following an arm's distance policy based on the lowest bidder, it is moving to establish a closer, longer-term relationship founded on a recognition of mutual stakes in quality and efficiency, in which it gives clear guidance and even offers training in quality control methods to some of its vendors. In other words, it is telling its vendors that Hewlett-Packard expects, and insists upon, high standards of production.

In Japan, that sort of long-term relationship has been characteristic of the postwar structure of subcontracting; it gives the Japanese companies a measure of flexibility, resiliency, and cost competitiveness, not to mention quality workmanship.[12] By paying closer attention to production technology, and by being willing to spend the capital necessary to ensure high reliability, American firms are rising to the Japanese challenge and are exposing the shallowness of stereotypes about inferior American manufacturing quality. Hewlett-Packard and Advanced Micro Devices, among others, have already established reputations for quality workmanship. Competitive pressures are forcing other U.S. companies to follow suit, though how successfully and far the stress on process technology spreads is not yet clear. At least one indicator, capital investments, suggests that the trend may continue.

Capital investments. Capital investments, the other prime catalyst for innovation, and market demand are closely related. If market demand provides the incentives, capital investments supply the resources and tools for innovation. Investment in new plants and facilities, as pointed out earlier, is a crucial component of process technology; R&D investments supply the lifeblood for new products.

Capital investments are broadly correlated to a host of dependent variables including market share, cost and price reductions, product quality, and reliability. In the fast-moving semiconductor industry, rates of capital investments, in both the U.S. and Japan, are exceptionally high and show no signs of slackening. Semiconductor companies realize that if they fall off the investment pace, they run the risk of being left behind, or of having to drop out of the race. From 1977 to 1980, U.S. firms invested on an average about 8 percent of sales for R&D, a figure nearly three times higher than that for other manufacturing industries. For technological leaders such as Intel and Texas Instruments, the ratio of R&D investments to sales has been even higher. Intel, whose sales revenues in 1982 came to $900 million, allocated $130 million for R&D in 1983. The range between 12-15 percent is by no means out of line with what is considered necessary to maintain technological leadership. Such high levels reflect the rising costs and uncertainties of state-of-the-art R&D, and the ferocity of competition among firms of rough technological parity.

As Chapter 5 showed, Japanese semiconductor companies are plowing back capital into R&D at a rate reportedly as high as 15-17 percent of sales revenue. In actual amounts, the U.S. is still investing in multiples of Japan, however. Indeed, two U.S. firms, Texas Instruments and Motorola, invested more in 1979 and 1980 than the whole Japanese semiconductor industry. This suggests that U.S. industry is in a very strong position with respect to R&D investments, just as it is in terms of market demand.

One can expect Japanese companies to sustain high levels of investments in order not to fall too far behind their American competitors. In other sectors Japan has won world renown for its pace-setting rate of capital investments. It is one reason why, in one industry after another, the Japanese have displaced American leaders.[13] Investing heavily over time is a logical strategy for Japanese companies to take—not simply because it is a proven formula for industrial catch-up but also because it takes advantage of Japan's extraordinary record of capital formation and its unique structure of financial institutions through which capital is distributed. The average savings ratio for households in Japan is around 20-22 percent of disposable income.[14] Whatever the reasons for this, and many have been cited, it has provided the underpinnings for Japan's system of indirect bank financing. The channeling of household savings through banks means that a large supply of capital is available to companies for investment purposes.

America's tax provisions, interacting negatively with high inflation rates, have had the effect of dampening investments. This means that macroeconomic management is an important determinant of investments. Here too, Japan has done very well—better than America since the mid-seventies. If higher growth, lower inflation, and lower interest rates can be sustained, Japan's semiconductor industry may be able to close the investment gap more speedily than Americans anticipate.

In spite of the huge quantity of R&D money being invested in the U.S., many American semiconductor executives believe that the aggregate figures hide an enormous amount of wasteful duplication. Some degree of duplication between firms is inevitable and indeed desirable from a competitive standpoint, but U.S. semiconductor spokesmen believe that joint work would improve the cost-effectiveness of aggregate R&D expenditures. To this end, they are advocating changes in antitrust laws. Another attempt to reduce wasteful duplication is evident in a new company, Microelectronics and Computer Technology Corporation (MCC), formed in 1983 through joint sponsorship of ten companies for the purpose of helping U.S. firms pool and coordinate R&D resources in order to respond more effectively to the Japanese challenge. Key areas of technology targeted for MCC research include advanced computer architecture (especially artificial intelligence-based expert systems), CAD/CAM, VLSI, and software development.

Innovation and employment practices. If market demand and R&D investments are moving in positive directions, are Japanese institutions capable of converting such external stimuli into innovative responses? Are such time-tested practices as lifetime employment and seniority promotion inimical to product innovation? Is the virtual absence of a whole stratum of merchant houses a serious handicap? Are pressures to innovate forcing large, vertically integrated firms to adapt their organizational structures? Is Japanese industrial structure itself a stumbling block?

The effects of such practices as lifetime employment and promotion and compensation by seniority are mixed. Facile generalizations fail to capture the complexity of their impact on innovation. On the negative side, incentives to innovate can be stifled when age and experience instead of individual performance serve as the standards for promotion, compensation, job assignment, and level of responsibility. Gifted young researchers in the prime of their creative years are sometimes said to be frustrated by the rigidities of Japan's hierarchical organization. Japanese management, aware of the problem, is trying hard to create a more flexible research environment that gives fuller play to the talents of young scientists and engineers; but Leo Esaki, a Japanese Nobel Prize winner who won fame for his study of the tunnel effects of electrons, is the best-known scientist who left Japan to work abroad in a research environment more to his liking.

Permanent employment can also be inhibiting. Leo Esaki has observed that "Low mobility is a major obstacle to creative thinking in Japan."[15] It limits the range of research specialists with whom there can be fruitful interaction. To avoid possible internal stagnation, Japanese companies must continually infuse their research staffs with new blood drawn from college and graduate schools. Moreover, as mentioned in Chapter 4, the low level of labor turnover may slow down the diffusion of state-of-the-art technology among semiconductor firms. Japanese corporations do not have the leeway their American counterparts have—to gain immediate access to cutting edge technology by hiring away specialists from competitor companies. Technological diffusion within the semiconductor industry may be less rapid as a result.

The research environment in the U.S. seems, on the surface at least, to be better suited for new product design and innovation. Skilled personnel move fluidly between firms in response to commercial and technological opportunities. The structure of the U.S. semiconductor industry is highly differentiated in sizes and types of firms (i.e., small, merchant houses and large, diversified systems corporations). Moreover, personnel practices within firms are seldom encumbered by seniority considerations. U.S. corporations can hire on the basis of specific needs and specialized skills, and they can pay, promote, or fire strictly on the basis of employee merit and achievement; they do not have to retain "dead wood" simply because of a commitment made at the time of hiring. Either the employees produce or they leave. High labor mobility is of course not without costs to individual firms, which must continually adjust to turnovers in personnel; but from the standpoint of the semiconductor industry as a whole, the system works to accelerate the overall pace of technological development. Mobility is an effective hedge against the kind of rigor mortis that occurs when companies get "locked into" established technologies because they want to extract profits from old product lines as long as possible.

On the other hand, lifetime employment is not inimical to innovation in all ways. It functions, for example, to hold down spiraling salaries for skilled research talent in circumstances where market demand exceeds manpower supply. Japanese bankers express astonishment at the exorbitant salaries (with generous stock options) paid semiconductor executives and first-rate research personnel.[16] Because of the system of permanent employment and seniority promotion and compensation, Japanese companies do not have to offer such "outlandish" incentives and rewards to retain key personnel. If American salaries keep rising in response to labor market conditions,* will the U.S. semiconductor industry be able to stay cost competitive? Many U.S. companies already feel stretched to the limit; will the spiraling costs of

---

* The American Electronics Association has projected a serious shortfall in electrical and computer-science engineers from 1984-89—somewhere around 113,500. If the laws of supply and demand operate, this will mean that salary pressures will continue escalating upwards.

skilled manpower place them at a deepening disadvantage? Will industrial salary structures continue to serve as disincentives for faculty recruitment and graduate school enrollment, the situation noted in Chapter 2? Here is an area, in short, where strongly held organizational principles (i.e., permanent employment) buffer Japanese corporations from the pulling and hauling of labor market forces felt fully in other industrial systems.[17]

Permanent employment may offer advantages in yet another sense, namely in terms of the continuity and collective experience gained when a fairly stable core of scientists and engineers work together over a long period of time within a multidisciplinary framework. Lifetime employment encourages long-term investments in manpower training and promotes close ongoing flows of communication among workers. America's mobile labor force almost guarantees a cross-pollination of ideas and allows individual companies to respond swiftly to new technological developments and market opportunities. But high labor turnover exacts a high price in terms of information loss, transaction costs, potential legal entanglements, discontinuities in experience, loss of company investments in manpower training, and the slowing down of momentum. In an era of VLSI, which seems to require teamwork and effective resource mobilization, ongoing and stable interactions among workers can be conducive to complex problem-solving.[18]

Innovation and industrial structure. Perhaps a more serious stumbling block to innovation than the seniority system or permanent employment is the almost complete absence in Japan of small-to-medium merchant houses, so central to the history of semiconductor invention. For reasons that require further research to pinpoint precisely—such as organizational flexibility, entrepreneurial incentives and rewards, weaker commitment to old product lines, and faster "turn around" time—small-medium semiconductor companies in America have accounted for a disproportionate number of major inventions following the basic groundwork laid by Bell Labs.[19]

The relationship between company size and innovation is controversial. Some studies show that only large companies are capable of mobilizing the money, manpower, and production base necessary to innovate effectively over time.[20] Other empirical studies reveal, however, that the rate of innovation is definitely higher for small-medium companies.[21] The diametrically opposed conclusions reached may be due to differences in the specific industries studied and in the definitions of innovation used. One study draws a distinction between industries with technologically "fluid" product lines and those with standardized product lines and limited leeway for technological change.[22] For mature industries where competition centers on production costs and prices, process innovation is emphasized and large size is an advantage. For technologically fluid industries, the emphasis is on new product design and performance, and small size seems to be an asset. The semiconductor industry falls into the latter category, and the contribution of the small merchant houses to semiconductor innovation would seem to confirm the validity of the hypothesis.

If small size is conducive to innovation, is Japan's industrial structure a major barrier? The answer depends, in part, on two factors: (1) the responsiveness of big firms to market and technological opportunities, and (2) the possibility of establishing in Japan small enterprises, or their functional equivalents, on the model of venture capital firms in Silicon Valley. With respect to the first, it should be pointed out that large firm size certainly does not preclude the possibilities for innovation. The seminal breakthroughs in semiconductor technology were first made at Bell Labs, and one does not hear many criticisms being directed at IBM or Texas Instruments, two large corporations, for failing to sustain an atmosphere conducive to innovative R&D. There is no reason why large size, by itself, should stop Japanese firms from demonstrating that they, too, can innovate. Also, of course, the way a company is organized is more important than size per se.[23] Well-organized big companies outperform poorly organized small ones. It is true that small, spin-off companies in America are credited with many major semiconductor inventions, but it may be that in some cases the actual work for these inventions was done earlier in large companies, from which smaller ones spun off. When labor turnover is high, attribution of credit is sometimes difficult.

Few definitive statements, therefore, can be made.[24] Both small and large firms have contributed to the history of semiconductor innovation. Small firms may have contributed more to components development since the mid-sixties, because of greater organizational flexibility, stronger incentives to come up with product design, and quicker turn-around time in exploiting emerging opportunities. But big firms have not necessarily been held back by size. Indeed, following Schumpeter and Galbraith, one might even argue that trends within the semiconductor industry—namely, escalating capital costs, greater research uncertainties, and shorter product cycles—place big firms in an increasingly advantageous position to carry on R&D. Small firms may be hard pressed to keep up with spiraling capital requirements.

If the relationship between size and innovation is complex, perhaps the most that can be said is that an industrial structure that combines both large and small firms offers advantages over a structure that is made up of only one type. In this sense Japan's semiconductor industry might benefit from the creation of a new tier of small entrepreneurial firms similar to those that exist in America, and that venture capital is continually bringing into existence. The availability of venture capital encourages not only entrepreneurial risk-taking but also new product differentiation, which small, start-up companies consciously adopt as a strategy to establish a niche in the marketplace.

As Chapter 5 showed, the absence of a venture capital market in Japan makes it very difficult to follow the Silicon Valley model. Venture capital permits entrepreneurs with creative new ideas to develop them into new products by providing initial financial backing.[25] Once the enterprise succeeds (and the failure rate is very high), the venture firm can go public by offering stocks and bonds. Intel, Advanced Micro Devices, and Mostek are just a few of the successful firms that followed this path. Over a billion dollars a year is available in the U.S. for venture investments. The cumulative total, as of 1983, exceeded $7.5 billion. Two conditions—the lowering of the capital gains tax, and the high rate of return on venture investments—have stimulated a substantial growth in this market. From 1976 to 1981, the annual rate of return exceeded 50 percent, compared with Standard & Poor's 400 Stock Index of 14 percent.[26]

In Japan, as of 1982, cumulative venture investments came to less than $90 million. Those Japanese entrepreneurs bold enough to gamble on leaving the security of lifetime employment usually have to draw upon their own resources to start a new enterprise. Banks are not normally willing to back new high-risk endeavors; in fact, some banks may not be willing to extend loans, even after small firms have demonstrated their viability. As a result, 90 percent of all start-up capital in Japan comes from the entrepreneur's personal funds or from relatives and friends.[27]

The virtual absence of venture capital in Japan may seem puzzling in view of Japan's advanced technology and rich pool of human resources.* But there are major impediments to its development. Japan still has no stock market vehicle designed to handle small start-up companies. The requirements for listing on the OTC (over-the-

---

* A few venture companies have been successfully formed and are contributing to innovation; for example, Kangyo Denki Kikai developed a noncontact electronic displacement sensor that caused a stir in sensor technology. *Japan Economic Journal*, Nov. 30, 1982, p. 16.

counter) market are currently so stringent that the securities of only
111 companies are being traded, compared with 13,000 in the U.S.[28]
Without a vehicle for easy entry into the stock market, venture capitalists find liquidation of venture investments difficult.

MITI is well aware of the problem. It is trying to make institutional
changes that will permit small start-up companies to list their equities
by lowering capitalization requirements. It is also thinking about tax
deferrals and better protection against bankruptcy for investors. In
this era of high technology, MITI feels that a venture-based system for
company start-ups would be very desirable. Japan would have an
ample supply of capital available, both at home and from abroad, if
institutional bottlenecks can be overcome.

Suppose that supply-side problems can be resolved. Is there apt to
be buoyant demand for venture capital? Chapter 5 suggests that labor
market factors may be inhibiting. Giving up the security of lifetime
employment is a very big risk to take: reemployment is costly and uncertain if the start-up enterprise fails. Even if the enterprise is successfully launched, the recruitment of top-quality engineers, scientists,
and programmers is likely to pose problems.* In Japan, unlike America, the entrepreneur who leaves a company to start his own firm is not
held in particularly high esteem; in fact, some stigma may even be attached to his defection. Hence, the bottlenecks today are not trivial.

In the near future, the structure of Japan's semiconductor industry
will probably remain unchanged. The top ten companies will continue
to dominate the industry, even though their rank-orderings may change
from time to time. This is not to say that there is no room for small
firms. They are, in fact, being organized all the time, though primarily
as offshoots of large parent companies. It is not altogether out of the
question that these subsidiaries will someday serve the same basic
functions as venture firms in America. Parent companies can provide
generous financial backing, select promising researchers to join the
firm, and give it lots of leeway to function independently. Some semiconductor executives interviewed expressed hope that small spin-off
subsidiaries might compensate for the rigidities of Japan's industrial
structure, especially in such fields as software development, which
seems to flourish within small organizational settings.

But the Japanese subsidiary may not be an answer to the structural

---

* On the other hand, if venture capital becomes widely available and start-up firms
begin springing up everywhere, the impact on permanent employment, rates of labor
mobility, and seniority compensation could conceivably be significant. Over time,
Japan's industrial structure might even be transformed. All this is, of course, pure specu-
lation. No one knows to what extent a venture capital market will take hold, much less
what impact an outcropping of small venture firms might have.

problem of how to organize on a small scale for innovation in an industry already dominated by big companies. Such subsidiaries do not
offer, for example, the same set of incentives and rewards as venture
companies in the U.S. And the reason they are being organized has less
to do with improving the environment for innovation than with gain-
ing flexibility in personnel assignment and diversifying risks.[29] If subsidiaries become a refuge for parent company employees on the verge
of retirement, or for those only marginally productive, the whole idea
of creating the functional equivalent of venture firms will turn out to
be illusory.

Experimenting with small start-up companies is not restricted to big
Japanese corporations. Two well-known American companies, Intel
and Shugart, have created their own venture-type groups. Intel Mag-
netics was started to develop bubble memory technology and Shugart
Microfloppy Venture Group was organized to produce small, $3\frac{1}{2}$-inch
microfloppy disk drives. The Shugart team is composed of about 50
employees, who have been given considerable leeway to operate, free
from the encumbrances of bureaucratic red-tape. Its members have also
been given attractive financial inducements, including cash bonuses for
meeting performance deadlines and some percentage of sales revenues—in addition to their regular salaries. The group has the advantage, which wholly independent start-ups usually lack, of having
direct access to Shugart's resources, including its sales and distribution
networks. Although, inevitably, a variety of problems have to be
ironed out, the idea of creating small venture firms within established
corporate structures is one that may merit further experimentation,
particularly by the Japanese.

It should be noted that, unlike America, Japan's small-medium enterprise tier has never been a seedbed of technological innovation. A
National Science Foundation study that analyzed the relationship between small-medium firms and the number of innovations in five
countries found that Japan had the lowest correlation.[30] The rate of
R&D activity in Japan's small-medium sector is very low: only 8 percent report that they even engage in R&D, compared with 56 percent
among large companies.[31] Several reasons can be cited to explain this
situation—a striking contrast to the United States, the United Kingdom, and France—including the latecomer syndrome, shortage of
capital, and labor market conditions. But the conspicuous lack of
technological innovation in the small-medium sector may also be, in
part, a reflection of Japan's industrial policy, aimed at rapid catch-up
using the full-throttled horse power of big corporations.

The price of Japan's industrial policy, often overlooked in the lionization (or vituperation) it receives abroad, has been its mixed impact on what in other countries is very often a source of enormous technological dynamism: small-medium firms. In postwar Japan, large, established corporations have been given preferential treatment; they have been the beneficiaries of Japan Development Bank backing, priority borrowing status, participation in national research projects, special provisional development laws, and so on. Small-medium firms have had to live with such hardships as higher costs of capital, much weaker drawing power for the recruitment of top-notch manpower, less favorable access to government procurements, fewer direct benefits from government R&D, and much higher bankruptcy rates.

Of course, MITI did not deliberately design the system to discriminate against small-medium enterprises. Some of the disadvantages mentioned above come with being small. Others are unintended consequences of an industrial policy that has tended to favor bigness. The systematic bias toward big business is the result of priorities placed on efficiency, scale economies, resource concentration, technological catch-up, learning curve, export prowess, and international competitiveness. To alleviate the difficulties under which small-medium firms must operate, MITI has devised a range of programs to assist small-medium enterprise—special financial institutions for low interest loans, special tax provisions, information and management counseling services, export facilitation, and technical training and assistance. The Small-Medium Enterprise Agency claims the largest budget within MITI each year.[32] Nonetheless, MITI's programs, however helpful, cannot fully redress the handicaps under which small-medium enterprises operate.

In particular, MITI's efforts to raise technological levels of the small-medium sector have fallen far short of expectations—in spite of the fact that this has been a basic objective since the early 1950's. Small-medium enterprises receive few R&D subsidies from MITI. The bulk of the money from the Agency for Industrial Science and Technology, Japan Research and Development Corporation, and Japan Development Bank go to big business—for the simple reason that the best research personnel, facilities, and support capabilities are concentrated there. Small-medium companies have access to two funding sources specifically designated for them: the "Technical Development Subsidies to Small and Medium Enterprises" and the "New Techniques of Industrialization" fund, managed by the Small Business Finance Corporation. But the total amount of capital available is very small; in 1979 the two funds provided less than $30 million, a proverbial drop in the bucket. The network of 195 laboratories scattered throughout Japan's forty-seven prefectures is supposed to carry on research of direct interest and relevance to small-medium enterprise. But this prefecture-centered system, the core of the government's program, has failed to lift levels of technology significantly, much less turn the small-medium sector into a fertile seedbed of innovation. The government does provide moderate tax incentives for R&D, incentives for technological exports, loan programs for the purchase of new equipment, and lower corporate income tax rates for small-medium firms. Some evidence of progress is discernible, thanks to such efforts, but it will be some time before small-medium firms in the semiconductor industry function on a technological par with venture firms in America.

Japan's industrial structure poses possibly another disadvantage. It stems from the fact that most of the leading Japanese semiconductor manufacturers are large, vertically integrated and diversified firms. Though this means that large Japanese firms enjoy certain financial advantages, as Chapter 5 points out, it also may have exacted costs in terms of technological development. In large, vertically integrated and diversified firms, semiconductor divisions tend to be dominated by stringent design, performance, and manufacturing specifications set forth by systems divisions.

Certain Japanese firms have tried to work around this problem by creating, in effect, two semiconductor divisions, one for systems applications and the other for components development. This organizational adaptation is designed to reduce the technical costs of vertical integration, but how successful it has been is not yet known. The history of semiconductor invention suggests that the autonomy enjoyed by specialized merchant houses is one reason for their innovative ingenuity.

If vertical integration poses problems for semiconductor components development, the trend toward mergers and capital acquisition of U.S. merchant houses by large systems corporations is perhaps not an auspicious development. The circumstances giving rise to the trend are perfectly understandable: mergers and outside takeovers are structural responses to such pressing problems as escalating capital intensity, dwindling profit margins, high interest rates, and an economy caught in the trough of a deep recession. How it affects the development of semiconductor components within acquired merchant houses will bear close observation over time.

The agreement between IBM and Intel (1983), whereby IBM gained a 12 percent equity interest in Intel for $250 million cash, is a particularly intriguing development in that it stopped short of merger (or take-over) and yet relieved Intel of immediate worries about cash flow. With the $250 million, Intel reportedly plans to invest in new plant and equipment, repay short-term debts, and meet other needs. In a similar but separate transaction, IBM acquired a 15 percent equity position in the Rolm Corporation, known for its capabilities in tele- communications technology. IBM and Rolm plan to cooperate in devel- oping new equipment, perhaps in areas that link telephone networks and office and home automation. From the perspective of Japanese companies, which are unable to complete mergers and acquisitions as easily (in part because of the barriers posed by permanent employ- ment), such developments appear to give American firms enviable flex- ibility to strengthen vital areas of technological and commercial op- portunity. However, it is still too early to tell how such arrangements will work out. So far at least, the after-effects of IBM's maneuvers have not been entirely negative for Japanese companies. At least one Japa- nese firm reported a rise in purchase orders for such products as mi- croprocessors from American companies that do not want to be overly dependent on supplies from Intel; their fear, apparently, is that IBM's demand for semiconductor components is so big that Intel's output may wind up being pre-empted. To guard against the possibility of being caught in a shortfall, therefore, these companies have diversified their sources of semiconductor supplies.[33] Here, as elsewhere, the long-term effects of structural trends, now only dimly discernible, are complicated and exceedingly hard to anticipate.

Summary. In semiconductor technology, the pace of change is so fast that accurate forecasting is virtually impossible. The best that can be done is to assess the likely implications of trends in the two vari- ables—market demand and R&D investments—that appear to have the closest bearing on product and process technology. "Demand pull" and "investment push" factors can then be placed within the larger context of what might be called "conversion mechanisms" such as in- dustrial structure, company size, corporate organization, and employ- ment practices. A composite picture of trends, institutional strengths and weaknesses, and factors that might alter trajectories can then be pieced together.

The systemic approach taken in this study provides a framework for analyzing macrolevel factors, not normally subsumed under firm- specific analysis. It lays out the broad linkages between variables. From a macrolevel perspective, we can say that the U.S. industry ap- pears to be in a stronger position with respect to technology than its

Japanese counterpart. The U.S. industry set the pace and appears to have the capacity to maintain its lead, at least in the near-term. At present, U.S. firms account for over 60 percent of world semiconductor production, and the sheer volume and composition of end-user demand will be likely to continue pulling technology in ever more sophisticated directions. Of course, to repeat an earlier caveat: aggregate forces are merely national trends, and the whole outlook could change if, say, one company happens to hit upon a seminal invention that throws technological trajectories off course. But since one can hardly predict such developments, trends in market demand and investment rates still represent perhaps the two best (but admittedly rough) indicators of what lies ahead.

Chapter 5 notes that in terms of absolute figures, U.S. investments far exceed aggregate Japanese levels. The U.S. can be expected to continue spending more in absolute terms over the foreseeable future, even though Japan is narrowing the gap and can draw upon a variety of strengths inherent in its financial and civilian R&D systems. Japan's expanding share of the world market now accounts for nearly 25 percent. As Chapter 3 showed, Japan is at the frontiers in mass-produced memory chips, optoelectronics, gallium arsenide, silicon-on-sapphire, high frequency transistors, liquid crystals, and light-emitting displays. America continues to hold the lead in such areas as complex logic systems, microprocessors, and software applications. From a market demand standpoint, the outlook is for stepped-up competition in all areas of technology: products, processes, and software. To do well, American and Japanese industries will have to demonstrate strengths in all three areas. Aggregate trends indicate that both industries ought to be able to continue making significant technological and commercial progress.

Financial Trends

In addition to technology, which is fast moving and therefore hard to forecast, competitiveness in semiconductors will hinge on conditions related to finances, not only at the company level but also in terms of national financial systems—such things as investment rates, pricing strategy, market share, profitability, and long-term staying power. The importance of financial factors is growing as the capital intensity of the semiconductor industry rises. Estimates cited in Chapter 5 place the ratio of capital expenditures to sales somewhere around 0.16-0.17 in 1980, and climbing. These rising capital intensity levels are due to a variety of general conditions: escalating costs of new plant and equipment; faster rates of equipment obsolescence; higher costs of skilled labor; contracting product life cycles; rapid product replication and second-sourcing; shorter periods, therefore, within which first-to-market firms can recoup "up-front" investments; greater costs, risks, and uncertainties associated with state-of-the-art R&D; intensified price competition; and stepped-up competition in all areas, especially because of Japan's rapid emergence.

In this section, several questions will be considered: Does Japan's financial system confer competitive advantages on its semiconductor firms? Are Japanese banks willing to make risky loans because of tacit government guarantees? Is capital more abundantly available and cheaper? Are Japanese companies less concerned about high return on investment? Are they able to pursue long-term strategies, not driven by the pressures of short-term profit? Answers to these and other questions can give us insights into the competitiveness and staying power of Japanese and U.S. semiconductor industries.

Alleged advantages. Japan's financial system, featuring extensive bank lending to corporations, is alleged to give Japanese companies a variety of advantages. Access to capital is said to be easier, faster, steadier, and cheaper. Banks appear willing to lend to companies under seemingly risky circumstances that in America would never be considered possible, such as companies with inordinately high debt-to-equity ratios or firms in dire straits. The presumption is that the whole system works because the government stands behind it, ready to redeem bad loans or bail out failing companies and banks. One study has stated succinctly: "At bottom, then, despite the risks of high leverage, the resulting system is stable because government concern with the well-being of firms in favored sectors, like semiconductors, is taken as an implicit guarantee of loans made to them." [34]

Debt financing in Japan is also said to be cheaper than equity financing in the United States. Japan's high level of debt financing means that Japanese corporations can worry less about return on investment (ROI), quarterly profits, price-earning ratios, and the company's stock value. It is free to pursue a long-term strategy aimed at expanding market share. Indeed, according to Peter Drucker, Japanese firms are driven to seek bigger market shares by the imperatives of high debt-to-equity financing (with 80 percent debt and 20 percent equity not uncommon) and lifetime employment. [35] The emphasis on market share over profit maximization is a winning strategy for industries like semiconductors that reflect the pronounced effects of learning by doing, and Japanese corporations are therefore said to enjoy major advantages over their American competitors, thanks to the country's unusual financial system.

Capital availability. Chapter 5 suggests that the close and complex relationship between banks and companies, which they call the "banking-industrial complex," confers important advantages on Japanese semiconductor companies in terms of the timing, circumstances, and amount of debt capital available. Raising capital from banks is simpler, is faster, is based on better information, and involves lower transaction costs than trying to raise it from an impersonal equities market. American companies have to worry about the timing of their new equity issues, since they want to be sure to offer them when prices are high. Furthermore, because Japan's leading semiconductor manufacturers are large, diversified firms, they tend to enjoy higher credit ratings and larger amounts of internal funds than small U.S. merchant houses.

Unlike commercial banks in America, Japanese banks can hold equity (of up to 10 percent until 1987, 5 percent thereafter) in corporations with which they do business. Convoluted patterns of overlapping and cross-cutting ownership further deepen the relationship. In this context, bank loans can be viewed as a form of preferred stock; banks stand to lose much more than the unpaid portion of their debts if companies with which they have an equity relationship perform poorly or go under. It is an interlocking relationship of shared interests and mutual dependence, which both sides take pains to sustain. It is not the arm's length, distant relationship, legalistically defined by contract, that American companies are accustomed to.

The fact that semiconductor technology has been dubbed a national priority within the framework of Japan's industrial policy is undoubtedly advantageous in terms of access to capital. The semiconductor industry receives some preferential loans for technology development from public coffers. In 1980, the Japan Development Bank dispersed over $300 million to the electronics industry for computer-related development. These loans are offered at a percentage below commercial interest rates; but given bank practices of requiring compensating balances, the actual spread between preferential and commercial rates may be larger.* Moreover, government loans tend to communicate sig-

---

* The amount of compensating balances varies across firms and business conditions. Some of the very large and profitable corporations are not required to leave large balances, whereas other companies are forced to carry fairly heavy ones. As the Japanese government loses its administrative authority to set interest rates, however, the need to rely on compensating balances to adjust real interest rates has diminished.

nals to private banks, which look favorably on "chosen" sectors. Loans are considered sound investments, because priority industries are usually the high-growth sectors for Japan's economic future. Although the government does not guarantee loans, its designation of the semiconductor industry as a national priority does communicate a message that it will do what it can to facilitate the industry's development.

There is no shortage of capital available in America for its semiconductor industry, in spite of the absence of an industrial policy that confers preferential status. Capital markets have done a satisfactory job of funneling funds, thanks to the industry's past performance, growth potential, and competitiveness. Where U.S. firms might be at a disadvantage vis-à-vis Japanese competition is in the flexibility of capital access allowed by bank-firm relations. The long-term and interdependent nature of that relationship in Japan provides better protection against downside risks. Japanese companies can usually count on bank support above and beyond contract obligations during periods when business is bad.

Of course, Japanese companies may need greater insulation precisely because their heavy burden of debt servicing, comparatively low profit margins, and high fixed costs of lifetime employment make them more vulnerable to fluctuations in business cycles. Bank support is certainly not a guarantee of a continuous flow of funds, and Japanese companies must pay a price for downside protection. Nevertheless, the long-term commitment of banks and companies to one another and the adaptability to changing circumstances made possible by that relationship have to be considered, on balance, net assets. Just how much of an asset depends in part on future patterns of growth and cyclical fluctuations in world semiconductor demand.

Debt financing: Leverage, risk, and government guarantees? Complex equity holdings, mutual dependence, and the availability of extensive information go far toward explaining why lead banks are willing to lend to keiretsu-affiliated corporations under circumstances that would cause U.S. banks to back away. One need not fall back on speculative theories of government backstopping. The nature of the banking-industry relationship itself is explanation enough.

The very idea that the Japanese government backstops the entire system implies that debt-laden corporations are poor risks. In certain circumstances (e.g., prolonged recession), there is some truth to that; but like other aspects of Japan's political economy, it is often exaggerated. Japanese debt/equity ratios have to be understood in context, not taken at face value. In Japan (as in the U.S.) corporate assets—

plant, equipment, offices, land, buildings, and so on—are not adjusted each year to take account of inflation, and therefore company assets are significantly undervalued. Japanese banks tend to view the current (i.e., inflation-adjusted) value of corporate assets as important sources of collateral. Its existence reduces the perceived risks associated with extending loans to highly levered firms.

Japanese accounting practices also overestimate corporate debts. Included here are compensating balances that banks hold as deposits against loans extended; it is their way of adjusting nominal interest rates. The amount of money set aside as compensating deposits varies by company, by loan, and across cyclical and monetary conditions. It is recorded as a company debt, even though it is held as a bank deposit.

Corporate debts also appear large because Japanese firms do a lot of business on credit. Large corporations regularly extend large amounts of credit to subcontractors and subsidiaries. At any given time, a typical company's category of accounts payable and receivable is large—roughly twice that of U.S. firms, reflecting the closeness of parent-subsidiary/subcontractor relations and horizontal ties between firms in the same keiretsu.

Also included under corporate debt are nontaxable reserve funds for retirement compensation and special contingencies.* In 1981 such reserve funds amounted to nearly 5 percent of total corporate assets for nonfinancial companies in Japan. The retirement reserve fund alone represented 16 percent of stockholder equity; in other words, employees at large corporations contribute de facto about one-sixth of equity capital. Reserve funds actually enhance the liquidity of Japanese corporations because management can draw on them for immediate needs, yet, they are recorded as debt. When adjustments are made for all these factors, the debt/equity ratio comes out looking much less risky from the bank's point of view.[36]

In the semiconductor industry, debt/equity ratios vary widely from company to company. At one extreme you find a company like Matsushita, operating on a nearly 100 percent equity basis; at the other extreme are companies like NEC that carry substantial debt. For firms falling into the latter category, no amount of juggling will alter the basic fact that they are more heavily levered than their U.S. counterparts. Nevertheless, the adjusted debt/equity estimates presented here place Japanese corporate finances in more realistic perspective. They help to correct distorted notions of government-guaranteed (or irra-

* In October 1982, special contingency reserves were shifted to the equity category.

tional) bank behavior. The calculus of risk is further reduced if, as Chapter 5 argues, some portion of bank debt is considered as a kind of "preferred stock." Seen in this light, the willingness of banks to extend capital to companies in which they hold equity interest is not hard to understand.

Over the years, Japan has developed an elaborate system for risk diffusion and reduction in the private sector. It begins at the bank level with the imposition of strict collateral requirements that have to be met by companies whose assets, as pointed out earlier, exceed listed estimates. The risk factor is further mitigated by the bank's power to intervene in managerial decisions. Under normal circumstances banks stay out of corporate operations; but if and when companies run into problems that threaten debt servicing and repayment, lead banks are prepared to intervene directly in management matters. Sometimes, depending on the severity of the crisis, lead banks may even assume direct control by dispatching their own people to run the troubled company until it is back on its feet.

Japanese banks are no less interested than foreign financial institutions in making sound investments. The notion that scant importance is attached to corporate profits so long as debts are being repaid is quite erroneous. Japanese banks assess long-term trends in profit earnings. If the trends are not positive and the outlook for improvement is bleak, there will be great reluctance to extend new loans. Japanese banks are not in the habit of handing out money indiscriminately; if anything, they may be overly conservative. There are, after all, alternative investment opportunities, and incentives are strong to channel capital into the best possible investment outlets.

Large Japanese corporations operate within a business environment that is structured in ways that diffuse and cushion risks. The mechanisms are evident almost everywhere: keiretsu groupings, subcontracting networks, parent company–subsidiary relations, trading company functions, industrial associations, and so on. Institutions in the private sector mesh with those in the public domain. Together, this private-public infrastructure of risk diffusion and reduction is one of the most striking features of Japan's political economy. It functions to permit greater risk taking in aggregate. Structures for risk reduction exist in virtually all countries, of course, and they often lead to unintended rigidities that undermine market incentives and allocative efficiency.[37] Japan's system is remarkable in that the dysfunctional side effects have been kept under relative control.

The existence of this system of risk reduction mitigates the alleged need for government guarantees. Neither MITI nor the Finance Ministry is in the business of guaranteeing loans or bailing companies out. One would be hard pressed to identify a single instance in which MITI has rescued a company on the brink of bankruptcy. The U.S. government has come to the aid of Lockheed and Chrysler, and numerous companies have been nationalized in Europe; but none come to mind in Japan. True, the Japanese government is committed to promoting the development of priority industries like semiconductors through the vehicle of industrial policy, but promoting a whole industry is not the same as guaranteeing loans or favoring single companies.

So far, at least, MITI has not had to decide whether to bail out failing companies whose survival might be deemed essential to the national interest. If the need should arise, the government could be prodded into taking emergency action. But such a possibility, or any extreme emergency such as a run on the banks, which would push Japan into taking swift action, is hardly distinctly Japanese. Other countries, including the United States, would no doubt act as quickly. What needs to be emphasized here is that neither the banks nor the major semiconductor producers in Japan or the United States operate on the assumption that the government will rush to the rescue. Theories of ultimate risk assumption by the state fail to explain bank-business behavior in either Japan or the U.S.

Costs of capital. A question of great concern to American semiconductor executives is whether or not the cost of capital is lower for Japanese semiconductor companies. The question is difficult to answer definitively. The most widely cited study, conducted by Chase Financial Policy, a division of the Chase Manhattan Bank, concludes that the cost of capital is "significantly" lower for Japanese firms because of their higher debt financing.[38] This may be true; but flaws in the Chase analysis leave the matter far from settled. Perhaps the study's most serious shortcoming is its failure to weigh industry variations *within* each country; the focus is almost exclusively on variations *between* countries. Without a systematic assessment of national variations, one cannot advance valid generalizations about cross-national differences. To put the problem in a slightly more technical cast: analysis of variance is the sum of between-group *and* within-group variations.

Another problem with the Chase study is that it mixes categories of analysis. It compares the finances of U.S. semiconductor firms with those of large, diversified Japanese corporations without breaking the data down into division units. A more accurate comparison would have used disaggregated semiconductor division data. The way the analysis now stands, there is no way of controlling for the effects of exogenous (i.e., non-semiconductor) factors. Of course, semiconductor division data may not have been readily available. In fairness to the Chase study, one should point out that collecting comparable data is no easy task. Even if seemingly comparable categories are found, subtle but significant differences in accounting procedures can introduce serious biases. Measuring differentials in capital costs is therefore complicated by practical constraints in the available data. Unless strictly comparable data are used, valid, cross-national conclusions simply cannot be drawn.

The Chase study is also vulnerable to criticism on grounds that it calculates the costs of capital in terms of abnormally high U.S. interest rates. The calculation should have been based on a weighted average of interest rates from 1970 to 1980; that would have provided a better yardstick of real costs over time. Similarly, there are problems in the estimation of equity costs, the treatment of yen-dollar exchange rates and exchange rate fluctuations. But for all its problems, the Chase study is one of the few attempts to measure what others have merely asserted (with scant empirical evidence). Although the measurement may be off, the conclusion itself—that Japan's cost of capital has been lower—is probably valid. To prove that it is, the calculations would have to be done again, using strictly comparable, appropriate, and accurate data.

The argument that Japan's cost of debt financing is lower than that of America's equity financing sounds plausible. Interest payments on borrowed capital can be deducted from taxable income, but dividends paid out to equity holders cannot. On the other hand, unless Japanese companies continue borrowing in order to pay back old loans, debt-servicing and repayment can impose sizable claims on current cash flows; dividend payments need not. And if debt financing is such an advantage in terms of obtaining cheaper capital, why are Japanese companies cutting back? As Chapter 5 notes, a trend toward less reliance on bank borrowing and more reliance on internal funding and equity/bond financing is clearly discernible since 1975. Does this mean that Japanese companies are deliberately choosing to pay more for capital? What lies behind the trend? Is it likely to continue?

The answers are not clear. Several possible explanations can be advanced, but research would be required to determine their validity. The time period 1975-79 happened to encompass an interlude of heavy foreign investment in Japan, largely as a result of diversifying petrodollar investments outside the U.S. Some Japanese companies were also granted permission to raise capital through the issuance of overseas bonds. How much foreign investment capital found its way into the hands of Japanese semiconductor producers is a question that only research can answer. If significant sums were raised, how much did that capital cost? Was it more or less expensive than debt financing? If less expensive, then studies of comparative capital costs must assess alternative capital sources *within* countries as well as between them. If more expensive, why did they pay more?

Probably the main reason for the shift toward greater equity financing is to be found in the internationalization of Japan's financial system. The lifting of old controls over capital movements means, in effect, that Japanese interest rates will be more volatile than in the past. Less predictable and higher interest rates will make it harder for heavily levered companies to cope. They will be exceptionally vulnerable to sudden and sharp surges in interest rates, owing to their combination of high fixed costs (associated with guaranteed employment) and low profit margins (associated with weaker ROE, return on equity, pressures). As protection against such an eventuality, highly levered Japanese companies are trying to reduce their degree of debt exposure. Both MITI and the Ministry of Finance apparently applaud the attempt, feeling that, in the face of truncated domestic control, Japanese corporations need to place their finances on a more evenly balanced footing.

Another plausible explanation, albeit of secondary importance, is that highly levered Japanese corporations want to rid themselves of the rigidities of bank dependence. Japanese management may feel, for example, that corporate autonomy would be strengthened by cutting back on debt; bank borrowing inevitably brings a degree of bank control, including the ultimate authority to assume temporary control over company management. Such incursions on company autonomy must be hard at times for managers to take. If lower capital costs and company independence exist in something like a trade-off, perhaps management prefers a healthier portion of independence, even if it means having to pay higher capital costs.

Whatever its other advantages, heavy debt financing seems to carry a high price tag in terms of non-capital costs. Few would dispute the contention, for example, that heavy debts pose potentially severe problems during business recessions. Even if debt is rolled over, companies must still cope with high fixed costs that cannot easily be cut back. If the recession is deep and prolonged, Japanese firms may find themselves staring in the face of bankruptcy. When given the chance, therefore, Japanese companies may have chosen to enhance their capacity to survive cyclical downturns by reducing their level of debt.

The trend toward lower debt financing in Japan thus brings out an important point: that there are potentially serious costs and risks associated with high debt/equity ratios, whatever the presumed advantages in terms of access to cheaper capital. The costs, risks, and benefits *vary over time* with cyclical changes in the business environment. Generalizations that apply at any given time may not necessarily apply when the conditions of competition change. Unexpected changes in semiconductor demand, technology, interest rates, and so on can turn the cost advantages of debt financing into liabilities.

The issue of capital costs goes far beyond comparisons between Japanese debt and American equity financing. To focus narrowly on methods of financing is to lose sight of the bigger picture. Ultimately, capital costs must be understood within the broader context of macroeconomics, especially trends in capital formation. If capital *is* cheaper in Japan, as many believe, the reasons undoubtedly lie in differential rates of capital formation. Japan has long had a much higher rate of savings than America, or any other country. What is amazing, and puzzling, from a theoretical point of view is the persistence of record savings in the face of fluctuations in inflation and interest rates, ups and downs in business cycles, and steep gains in disposable income (relating to the Engels coefficient). The marginal propensity to save in America seems much more sensitive to such factors.

Japan's pace-setting savings rate means that capital is plentiful. Indeed, since the mid-seventies, Japan has found itself in the awkward position of having an excess of savings over business investment demand. This situation—quite unusual in the world's economy—has given rise to capital cost advantages as well as serious fiscal and trade problems. To take up the slack in private demand, the government has had to resort to heavy deficit spending: in 1982 it exceeded 5 percent of the GNP (making it the highest in the industrial world). Difference in U.S. and Japanese patterns of savings also provide a monetary explanation for the huge bilateral imbalance in current accounts. Semiconductor trade, the yen-dollar exchange rate, price competitiveness, and the cost of capital are all affected by macroeconomic forces at work. One of the urgent tasks before policymakers in Washington, therefore, is to devise policies that stimulate significant gains in capital formation at home while at the same time coping abroad with the trade and financial implications of higher Japanese savings.

From a systemic standpoint, it may seem odd that national variations in capital supply should be reflected in capital cost differentials. If capital markets are fully open, the potential cost advantages of one country like Japan should be offset by massive capital movements (i.e., international arbitrage) or by automatic adjustments in yen exchange rates. But Japan's financial system has not yet been fully internationalized, though it is clearly moving in that direction; and the floating exchange rate mechanism has failed to function as expected in adjusting currency values to trade flows.* Here, in short, is an area of vital concern to the U.S. semiconductor industry, particularly if, as some observers claim, the Japanese government—for whatever reason—has been dragging its feet on the full liberalization of the country's financial system.

To assess questions of capital costs requires, in sum, a broad approach that takes capital formation and international capital markets fully into account. Mechanisms of capital allocation are important, to be sure, and carry implications for corporate management and strategy. But debt vs. equity financing is only one aspect of deeper macroeconomic forces at work.

Debt and equity financing: Implications. It is a widely held belief that indirect bank financing gives Japanese companies greater leeway to pursue long-term strategies in world competition. In the United States, reliance on equity financing is said to lead to a preoccupation with high returns on equity (ROE), strict monitoring and controls over investments, and an obsession with high stock valuation. Presumably, this inclines U.S. companies toward short-term goals and quarterly profit earnings that keep stockholders happy, and works against the profit-forsaking, market-share approach that is considered essential for staying power and competitiveness. By contrast, Japanese firms are said to come under much weaker return-on-investment (ROI) and stockholder pressures. Presumably, this permits them to invest heavily in new plant and equipment, price products aggressively, incur sustained losses, and expand long-term market share.

Data from the Chase study seem to substantiate these stereotypes. The study reveals that Japanese semiconductor producers operate on lower profit margins and lower capital turnover ratios than U.S. companies: in fact, Japanese semiconductor firms show an average rate of return on capital that falls below their cost of capital. U.S. firms also appear to fall slightly below the break-even point, though according to the Chase study their rates of return have roughly equaled their cost of

* Despite a huge trade surplus with the U.S. in 1982, the yen depreciated sharply relative to the dollar.

capital. If Japanese competition forces U.S. rates below the break-even point, the Chase study warns, this will damage stock prices and the industry's access to equity capital. Here again, Japan's financial system appears to bestow substantial competitive advantages on its semiconductor producers; but is this the case?

Return on investment. Chapter 5 does not delve into this subject in detail, but one of its co-authors, Hiroyuki Itami, has written a book, *Nihonteki Keiei-ron o Koete* (Beyond Theories of Japanese Management), that analyzes questions of capital utilization, comparing Japanese and American companies in three sectors.[39] Itami's findings also appear to substantiate important aspects of the foregoing portrayal, but the implications drawn are significantly different.

Itami finds that Japanese firms use capital less efficiently than U.S. companies in part because Japanese management is less concerned about earning a high return on investment (ROI).[40] The need to maximize short-term profit or to show hefty quarterly earnings is not a major concern. Not that Japanese companies disdain "filthy lucre"; in markets as competitive as Japan's, companies have to be concerned with profits. It is just that the system of indirect, external financing permits greater slack in the management of liquid assets.

Not surprisingly, the ratios of value-added to both sales and total assets fall substantially below those of U.S. firms.[41] The low level of value-added is largely a function of less complete vertical integration (with extensive networks of subsidiaries, subcontractors, and distributers) and higher price-to-cost ratios (i.e., prices closer to costs). The price-to-cost ratio reflects the intensity of competition in Japan, particularly where price, not product function or quality, determines consumer choices. Itami thinks that price competition is excessive because Japanese companies do not choose their product portfolios wisely. Contrary to stereotypes of exempiary planning, Japanese corporations seem somewhat deficient in systematic product strategies. Many appear to follow a one-set principle, producing one of everything simply because their competitors are. There is a haphazardness about the whole process, with little or no market analysis or strategic evaluation of how products fit into an overall portfolio. A better selection of products would raise value-added and relieve pricing pressures, Itami believes.

Why are Japanese managers, otherwise noted for their adaptability, so backward in the area of strategic planning, especially in view of the seemingly favorable leeway for the pursuit of long-term strategy made possible by debt financing? Itami links the portfolio problem to rigidities caused by Japan's lifetime employment system. Once facilities are built to produce certain goods, closing them down becomes difficult since the livelihood of permanent employees is at stake. An effective product strategy requires flexibility to retreat quickly from product lines of only marginal value-added.* Here again, we see that the complex institution of permanent employment brings with it advantages as well as drawbacks for Japan's industrial system.

With capital abundantly available, Japanese firms traditionally invest heavily in new plant and equipment as a means of improving product quality and lowering production costs. Such investments are mainly responsible for steep rises in labor productivity over the years. Companies compete intensely in the capital investment arena, because that extra "edge" acquired through labor-reducing mechanization can be converted into significant gains in the marketplace, and no one wants to be left behind. The availability of capital allows companies to enter the race and to stay in it over a sustained distance. One can expect this dynamic to continue driving Japan's semiconductor companies, just as it has other industries. Climbing capital intensity in the semiconductor industry seems to play to one of Japan's systemic strengths.

On the basis of an admittedly limited sample, not directly focused on the semiconductor industry (but including several semiconductor manufacturers), Itami provides evidence that seems to validate some of the commonly held notions about corporate financing in Japan: (1) capital is plentiful and relatively cheap; (2) bank borrowing relieves pressures for high ROI; (3) capital intensity is high but value-added is low; (4) U.S. firms outperform Japanese companies with respect to returns on assets (ROA) and capital productivity; (5) Japanese firms compete fiercely in capital investments; (6) products are priced closer to costs; (7) Japanese firms place less emphasis on stock market valuation; and (8) Japanese companies feel less compelled to maximize short-term profits and price-earnings ratios.

Implications. At first glance these conclusions appear to lend credence to the notion that Japanese companies take a farsighted approach to strategy, free of the short-term myopia caused by ROI imperatives. But neither Itami's book nor Flaherty and Itami in Chapter 5 of this volume offer evidence to confirm this view. America's capital market does not stop companies from following long-term, market-share strategies. Certainly no one is accusing American Microsystems, Inc.

---

* Of course, companies can be too quick to pull out of product markets. Persistence has paid off in many instances.

(AMI), Hewlett-Packard, or IBM of strategic myopia. Although pressures to maximize short-term profits and ROI are undoubtedly more strongly felt in American companies, and although such pressures can lead to an unhealthy preoccupation with short-term profits, the behavior of U.S. semiconductor manufacturers defies facile generalizations, particularly those based on what might be called "financial determinism." The strategy of a number of leading U.S. semiconductor companies would have to be described as long-term, adaptable, and market-share oriented. How else could one explain the high and sustained capital investments, extension into overseas markets, the commitment to quality control, the support for university-based research and manpower training, the bold new attempts to coordinate research activities, or the new programs for future generation technologies such as the Defense Department's Supercomputer Project? Short-term objectives (such as profits) need not be incompatible with long-term goals (such as market share). Strategic plans can be broken down into a series of objectives, the accomplishment of which in sequence advances the company ever closer to its long-term goals.

Nor does heavy debt financing necessarily force Japanese corporations to devise farsighted plans. Itami suggests that Japanese companies are not as adept as their American counterparts in managing money efficiently* or in conceptualizing and implementing strategic plans, at least in the area of product portfolio management. Others have also debunked the myth of Japanese management's "genius" for long-term planning. Kenichi Ohmae observes that the "stunning Japanese successes in the auto, semiconductor, and consumer electronics markets have taken place over the long haul, but have occurred primarily because of a determined focus on short-term, incremental gains."[42] The structural determinism underlying the view that dif-

---

* We cannot tell from Itami's data whether the annual financial statements of the Japanese and American companies in his sample were based on the same accounting procedures. If not, such national variations may account for his findings. Even if they were the same, companies can still hide or highlight figures in ways that give annual financial reports a certain slant. It is conceivable, for example, that some Japanese companies might actually want (under certain circumstances) to understate their profits—perhaps in order to discourage company labor unions from making wage demands that management considers unreasonable. By the same token, it is entirely possible that the financial statements of certain American companies betray a bias toward inflating yearly profits in order to keep credit ratings and stock values high. Since one cannot ascertain how much or little distortion may have been introduced by the raw data on which Itami's analysis rests, his conclusions must be treated as somewhat tentative. His conclusion that Japanese companies use capital resources inefficiently may be true, at least for his sample, but it appears to contradict other data that show Japan has the lowest capital/output ratio of any industrial economy.

ferences in corporate finance compel U.S. firms to follow a short-term strategy of profit maximization while Japanese firms pursue a long-term strategy of market share simplifies what is in fact a very complex set of circumstances on both sides.

This is not to deny that debt financing gives Japanese companies more "slack" to tolerate some sacrifice of short-term profit. There are noteworthy examples of a willingness to absorb early losses in order to break into a new product line or new industries—semiconductors being but one case in point. Such persistence is generally more difficult under stricter ROI guidelines.

Itami's study suggests perhaps that the system of debt financing inadvertently breeds inefficiency (in somewhat the same manner that military contracts and procurements almost guarantee some measure of waste in the U.S.). Japanese companies tend to forego the use of standard American methods of financial analysis—ROI, discounted cash flow, and so on—though they do try to ensure through regular monitoring that sufficient money is available before new investments are made.[43] Other costs that emerge out of Japan's structure of corporate financing include a penchant for overinvestment, excess capacity, excess competition, corporate vulnerabilities during recession, high bankruptcy rates, merger and capital acquisition difficulties, and the need for administrative guidance and government coordination. Whether the advantages of lower capital costs and more flexible capital utilization outweigh such disadvantages is not altogether clear.

In trying to understand and assess the costs and benefits of Japan's financial system, it would be well to remember the postwar context within which it evolved. Heavy debt financing grew out of an era of rapid growth and financial insulation—when Japan's equities and bond markets were underdeveloped. To finance their growth, Japanese companies had no choice but to rely on banks as their primary source of capital. Had the capital market offered a fuller range of instruments, the "banking-industrial complex" might not have developed in the way it has. But it did not, and the momentum of high-speed growth deepened bank-company ties and consolidated the system of external indirect financing. For years, the whole system was considered fragile, susceptible to collapse in the face of the first severe crisis. But full-speed growth allowed the system to operate effectively and with surprising stability.

The onset of a period of slow growth following the oil crisis altered the foundations of Japan's economy. Heavy debt financing no longer functioned as well as it had prior to 1973. It was, in many ways, ill-suited to an era of sluggish growth, high unemployment, and future uncertainty. The dramatic shift away from debt financing is no doubt a reflection of the transformation that has taken place in the wake of the oil shock. The trend seems likely to continue as long as growth rates stay sluggish. Some companies with very low debt financing—for example, Sony and Matsushita—are managing very well under harsh business conditions. The internationalization of Japan's financial system, as suggested earlier, will no doubt accelerate the trend toward greater equity financing. Thus, American and Japanese patterns of financing seem to be edging closer together, though it will be a long time, if ever, before there is convergence.

Both American and Japanese financial systems seem to be functioning satisfactorily—albeit in different ways—to meet the escalating needs of their semiconductor industries. Both industries enjoy ready access to capital, on comparatively favorable terms, mostly through internal financing but also from direct and indirect external sources. Japanese corporations can draw on a variety of strengths inherent in Japan's financial system, including sound macroeconomic policies, extraordinary rates of capital formation, close banking-business relations, and effective mechanisms of risk diffusion and reduction. Such formidable strengths in financial infrastructure are reflected in tangible advantages for Japanese corporations such as lower costs of capital, greater capital availability, lower transaction costs, and fuller information on which to base decisions. The nature of corporate financing in Japan also appears to ease the kinds of pressures to maximize short-term profits and ROI, which lie at the core of American management concerns.

This is not to say, however, that Japanese firms enjoy financial advantages of such magnitude that they will turn out to be decisive. There are costs associated with Japanese-style financing that have already been pointed out, such as the burdens of heavy debt-servicing, which render companies vulnerable to fluctuations in interest rates and downturns in the business cycle. Such costs are by no means insignificant. Nor should anyone lose sight of the fact that Japan's financial system is undergoing change—in response, especially, to the powerful forces of internationalization. If the economy is fully opened up, this will probably mean that some of the advantages that Japanese semiconductor companies have enjoyed—such as low and stable interest rates—will be diminished. In anticipation of such changes, a number of Japanese semiconductor manufacturers are acting already to decrease their dependence on bank lending and increase their proportion of equity financing. Just how far and quickly such trends proceed is a matter that bears close monitoring, particularly since financial considerations have become so important to the U.S.-Japan semiconductor competition.

The Role of the State

If the American companies hold a clear edge in technology and Japanese firms have advantages in finance, what can be said about the role of the state and the two political systems? Which side comes out ahead? Chapter 4 suggests that the nod clearly goes to Japan, but not for the reasons usually invoked—not, that is, as a consequence of the national conglomerate, "Japan, Inc.," or almighty MITI funneling huge subsidies to the semiconductor industry. Rather, Japan's advantages derive mainly from the structure of its political system. It is a system based on a very stable core of power, composed of the Liberal-Democratic Party's (LDP) control over the legislature and the economic bureaucracy's responsibility in economic policymaking. Out of this system have emerged public policies highly favorable to the promotion of business interests. This is a sharp contrast to the constantly changing, sometimes incoherent, economic policies of the U.S. government.

Close bureaucracy-business relations constitute another asset of incalculable value for Japan's semiconductor industry. The relationship ensures a heavy flow of information exchange, regular consultations, and usually consensus on broad technological objectives. Of course, the picture is not as efficient and smooth as that conveyed by "Japan, Inc.," in which companies function like autonomous divisions of a vast conglomerate, headquartered at MITI. Competition between firms is fierce. MITI is bound to have differences with various companies that may at times resent government interference. Tensions and problems are inevitable and require constant communication and compromise. However, the system, as a whole, functions far better than government-business relations in America, which are characterized by ambivalence, lack of mutual trust, and sometimes outright antagonisms.*

Still another of the great strengths of Japan's political system is the effectiveness of its central bureaucracy, especially the ministries in charge of economic policy. Nearly everyone who has observed Japan's government in action has marveled at the competence of its central administration. Here again, of course, the strengths of the government must not be overdone, lest the false impression be conveyed that

---

* An exception is the relationship between the farm community and the Department of Agriculture, which comes closer to the Japanese model.

Japan's public sector is a model of rational efficiency. The same kind of ailments that plague governments elsewhere are also evident in Tokyo: territorial squabbling between bureaucracies, organizational rigidities, and administrative inefficiency. The problems are especially trouble-some when issues fall between the boundaries of bureaucratic jurisdic-tion. Yet, public sector ailments in Japan seem somehow less severe than those that afflict the U.S. In terms of sheer size alone—of govern-ment, of the economy, and of the problems that perennially beset it—the U.S. must contend with circumstances much more difficult to con-trol. Moreover, as Chapter 4 points out, the power of the central bu-reaucracy in Japan means that management of the economy is less politicized, less vulnerable to the tugging of interest group lobbying.

This impression is substantiated by Japan's remarkable economic performance. Choose any indicator you will—growth rates, unem-ployment, inflation, investment, savings, or trade—Japan comes out consistently at or near the top. Who, besides the public, benefits? Japan's semiconductor industry. With high income elasticity of de-mand, the semiconductor industry is especially responsive to advances in aggregate growth rates. The faster the growth, the larger the de-mand for semiconductor products. If the Japanese government is able to sustain faster growth rates than the U.S. government, without trig-gering inflation, Japan's semiconductor industry is going to grow faster (other things being equal).

Other features of Japan's political system that give it such strength deserve mention: the priority given to economic objectives, the non-military orientation, more pragmatic application of antitrust, effective mechanisms for interest aggregation, enterprise unionism, and non-litigiousness. This adds up to a system that comes as close to being a businessman's paradise as any in the world. One need not resort, therefore, to stereotyped notions of "Japan, Inc." to explain the sup-portive environment in which Japan's semiconductor industry operates.

MITI. As the putative headquarters for "Japan, Inc.," MITI is often given the lion's share of credit for Japan's industrial successes.[44] How-ever, some scholars dispute the extent to which MITI should be given credit. Patrick and Rosovsky take the position that "the main impetus to growth has been private—business investment demand, private sav-ing, and industrious and skilled labor operating in a market-oriented environment of relative prices. Government intervention generally has tended (and intended) to accelerate trends already put in motion by private market forces."[45] The view taken here, and hinted at already, is that the importance of MITI's role has changed over time, and varies by sectors and phases of development. During the early period of post-war recovery and rapid growth, MITI played an indispensable role.[46] But that role has waned as Japan's economy has grown. Remember that MITI's extraordinary power during the fifties and early sixties derived from structural weaknesses in Japan's catch-up economy, which required central coordination and some measure of protection vis-à-vis the outside world. The powers it possessed used to include that of foreign investments regulation, technology licensing, infant industry protection, rationalization and recession cartel formation, and perhaps most important, foreign exchange rationing. The latter was particularly effective as a tacit "stick," since nearly all companies depended on foreign exchange to buy and sell abroad. MITI lost some of these powers with rapid development and the internationalization of Japan's economy.[47] Of course, MITI has found new grounds for asserting itself, as, for example, the need to mediate trade conflicts. But overall, its power has clearly ebbed since the early fifties.

This implies that a more differentiated view of MITI needs to be taken, one that takes into account the changes in MITI's power and disaggregates the industrial economy into sectors. MITI's influence is greatest and its ties closest with such capital-intensive, producer goods industries as steel, electrical power utilities, and petroleum. Even here, its power is considerably less than it used to be;* the same sectors are heavily regulated elsewhere and in some states even nationalized. MITI's influence is perhaps most restricted in certain areas of the consumer goods sector, such as watches and precision equipment; it is somewhere in between for intermediate goods like semiconductors. As pointed out in Chapter 4, MITI played a central role during the industry's early growth and through its catch-up phases. It has probably done a more cost-effective job of accelerating the growth of Japan's semiconductor industry than the U.S. government. Nevertheless, the primary driving force behind the industry's development is unquestionably market competition, not MITI intervention.

MITI's power to guide, coordinate, and if necessary, coerce the semiconductor industry is often exaggerated. One study, commissioned by SIA, for example, charges that MITI divides and allocates R&D, forms production cartels, implements a "targeted" industrial policy that distorts market forces and gives Japan unfair advantages, heavily subsidizes semiconductor R&D, issues directives blocking U.S. semiconductor imports, and encourages "Buy Japanese" leanings.[48] These

---

* Interviews with MITI officials reveal that differences and tensions arise even with the utilities companies, whose rates are set by MITI.

allegations convey the false impression that state planning and control, not the market mechanism, dictate the terms of semiconductor competition in Japan. But the industry's dynamics are such that it flourishes in highly competitive environments and languishes where market forces are weaker. It is no accident, accordingly, that the industry is most advanced in the U.S. and Japan, two of the world's most vibrant market economies, and not nearly so advanced in England or France, despite extensive state subsidies, import protection, and preferential procurements in both countries.[49] France's efforts to shore up the competitiveness of its electronics industry wound up perhaps weakening it in the long run.[50] This is the danger of trying to control competition: it often has the perverse effect of undermining the discipline of market incentives on which efficiency is based. France's failure to achieve international competitiveness through the standard instruments of industrial policy has led the state to take even more drastic action. In effect, it has nationalized the French electronics industry; but the requirements of semiconductor competition make it very doubtful that nationalization will be an ultimate remedy. The more market-oriented American and Japanese semiconductor industries will be hard to compete against.

Of the two market-oriented systems, the tradition of state intervention is historically stronger in Japan than in the U.S.[51] Washington is reluctant to tamper with the invisible hands of market forces except in those circumstances where cherished values like democracy, equity, and consumer rights are placed in jeopardy. Tokyo is less hesitant about stepping in to guide market forces if such collective interests as efficiency and international competitiveness can be advanced. One acts only in response to the malfunctioning of the market; the other takes more initiative to steer the market in desired directions. More than a century of industrial catch-up has given the Japanese government extensive experience in utilizing public policy to accelerate the growth of strategic industries.

Many of the criticisms currently directed against government tampering in Japan's marketplace have more to do with past practices, especially those of the fifties and sixties, than with present policies. For years prior to trade liberalization (until the mid-seventies), the Japanese government adopted trade policies that stacked the competition in its favor, especially strict controls on foreign investments and over international capital flows. Some of the very things that Americans complain about today—such as the encouragement of "Buy Japanese" practices, a consciously undervalued yen (which, in effect, functions as an export subsidy), and closed procurements—were prominent features of Japan's infant industry protection. Even as late as 1975, long after infant industry measures could possibly be justified, full liberalization had still not been completed. Although many of the most objectionable features have been abandoned under sustained foreign pressures, the advantages gained in the past have not been lost. Japan's semiconductor industry can compete head-to-head with American manufacturers today, in part because of the early momentum gathered behind protective walls. In this sense, complaints about "unfair" trade practices have some basis in the legacy of the immediate past.[52] But the rationale for infant industry protection is one that has been commonly used by latecomers, including the U.S. iron industry during the nineteenth century.[53] Though it certainly can be criticized from the standpoint of economic theory, infant industry protection is a generic problem commonly associated with latecomer status, not a phenomenon peculiar to Japan's semiconductor industry.

The crucial question, therefore, is whether unfair trade practices are currently being followed, now that the country's semiconductor industry has caught up. Not all the evidence is in yet; but the weight of it indicates that MITI practices are not out of line with what other industrial states are doing. Under international pressure, MITI has given up a variety of objectionable practices, including high import tariffs and export tax incentives; it has also opened national research projects (like the Fifth-Generation Computer Project) to foreign participation and has made patents from its VLSI project available to foreign companies. Compared to its defensive stance in 1975, MITI's behavior has changed significantly. Successive trade crises, combined with the maturation of Japan's semiconductor industry, have forced MITI to alter policies smacking of neo-mercantilism that served the country so well during an earlier era of very rapid catch-up. It now sees imports as lying as much in Japan's national interests as exports (because the threat of protectionism is often triggered by large imbalances in merchandise trade). A distinction therefore needs to be drawn between the past and the present MITI.

The thrust of U.S. government statements have singled out Japan for criticism, even though the French and British governments have intervened far more sweepingly in the marketplace. The Japanese feel they are being targeted for special and unfair malignment. If this is true, and U.S. officials would deny it, the obvious reason is that Japan's semiconductor industry poses a much greater threat to U.S. leadership than that of any of the European countries. If the Japanese engage in unfair trade practices, accordingly the impact on semiconductor competition carries far greater ramifications, making it much less tolerable than is true for weaker states.

Perhaps another reason Japan seems to be singled out is cultural differences and the Japanese lack of policymaking transparency. Japan's policymaking system is so different that non-Japanese find it very hard to comprehend. Not that it is secretive; it is just that so much transpires informally, often in the after-hours of work, that non-Japanese cannot monitor the process easily. The opaqueness is apt to breed suspicions when foreign goals and interests are thwarted. One semiconductor executive, testifying before the U.S. International Trade Commission, suggested that mysterious phone calls, believed to be from MITI officials, appeared to have led to the cancellation of orders placed by Japanese companies.[54] Informal patterns of consensus-formation make such accusations hard to prove one way or another.

The notion that MITI coordinates production and R&D cartels in the semiconductor industry is based on an exaggerated assessment of what MITI is capable of doing. Yes, MITI has tried to "rationalize" various industrial sectors in the past by encouraging mergers and if possible, an industry-wide division of labor, which would reduce wasteful duplication and excessive competition—but with decidedly mixed results. In most cases MITI has not been able to impose its best-laid plans for rationalization on the private sector. Nevertheless, the succession of failures has not dampened what is probably a deep-seated propensity to try to shape the parameters of competition in order to heighten efficiency and industrial strength. There is no lack of precedent—in steel, automobiles, and computers—that would give grounds to the suspicion that MITI might be trying to organize a broad division of labor within the semiconductor industry. Such suspicions gain reinforcement from the superficial impression that the electronics giants in Japan all have their designated niches: NEC, for example, specializes in telecommunications; Fujitsu in IBM-compatible mainframes; Matsushita in consumer durables; Hitachi in industrial equipment and power systems; and Toshiba in heavy duty electrical equipment. To place large diversified manufacturers in such one-dimensional categories, however, belies the fact that they compete fiercely across a spectrum of overlapping product areas: mainframes, mini- and microcomputers, home appliances, and so forth.

Competition in overlapping product areas suggests that each company produces a fairly complete line of semiconductor components, ranging from MOS memory ICs, CMOS logic chips, MOS microprocessors, custom MOS logic ICs, discrete devices, and so on.[55] There is no masterplan of product specialization, much less one of research complementarity. Is it realistic to think that MITI could actually orchestrate a neat division of R&D and production when companies are going all-out to outdo one another? Leaving aside violations of antitrust, would big, diversified corporations agree to confine themselves to narrow market niches when the nature of semiconductor technology and commercial imperatives seem to require that companies enter and stay in basic technologies and product lines? Far from it. Indeed, one Japanese semiconductor executive bemoaned the duplication as "excessive" and "wasteful" from an industry-wide point of view.[56] His comments sound very much like those of American semiconductor executives who favor some relaxation of antitrust laws to permit more joint research and the pruning back of wasteful R&D duplication.

There is no doubt that national research products organized under government auspices in Japan serve to reduce redundancy and achieve some R&D economies of scale. For national projects, the Japanese government does try to impose a rational division of R&D labor on participating firms. But such projects represent only a small portion of total R&D. Each company has its own agenda of research, which is carried out independently of official projects. Certainly there is no mechanism in place for MITI to divide up the research that takes place in the private sector.

Thus, MITI's power is more circumscribed than is commonly believed. It cannot but be circumscribed under the conditions that exist: strong private companies with a sense of their own interests, intense competition, large and complex industrial structures, an interdependent economy, and limited legal authority. Accordingly, as Chapter 4 points out, what it can do to enhance the competitiveness of Japan's semiconductor industry is largely limited to the following: (1) creation of a vigorous business environment; (2) facilitating consensus on technological objectives; (3) reducing the costs and risks of state-of-the-art research and development; (4) information gathering and analysis; (5) obtaining legal and tax incentives; (6) mediating trade and investment activities between Japan and other countries; and (7) designating semiconductor technology as a national priority within the framework of an integrated industrial policy. MITI is not in a position to generate much direct demand; nor can it guarantee technological innovation or private sector efficiency. It realizes that excessive government meddling can have precisely the opposite effect. It has therefore sought to work within, not outside, the framework of market forces.

Industrial policy. Since 1982, much ado has been made of Japan's industrial policy. U.S. leaders in government and industry have come to focus attention on Japanese industrial policy, believing it to be an underlying source of unfair advantage and trade friction. Public criticisms have been directed at MITI policies that according to American leaders have violated the letter and spirit of multilateral trade agreements. U.S. delegates have placed the issue on the agenda of OECD and GATT meetings, and American and Japanese officials (representing USTR, Commerce, and MITI) have devoted considerable time to the subject at bilateral working sessions. The SIA has played an especially active role in bringing the issue of Japanese industrial policy to the fore of public attention.

The SIA report on the effects of government targeting, cited earlier, charges that the Japanese government has funneled large R&D subsidies to the semiconductor industry, in violation of the Subsidies Code negotiated during the Tokyo Rounds. It also claims that industrial policy in Japan serves to reduce risks that permit companies to sustain high levels of capital investments and to price aggressively— perhaps engage in dumping or even predatory pricing.[57] Such practices, the SIA report asserts, provide the basis for a GATT Article XXIII complaint concerning anticompetitive behavior and subsidization.[58]

There is little doubt that Japanese industrial policy has accelerated the growth of Japan's "targeted" industries. Priority industries have in the past received preferential access to capital. Whether, in the absence of an explicit industrial policy, as much investment capital would have found its way to these industries is a controversial question. Probably not. Chapter 4 points out that a significant portion of capital funds during the early fifties came from the government's Fiscal Investment and Loan Program; between 1952 and 1955, FILP funds accounted for over 28 percent of total capital available to industry. The figure fell to around 15 percent between 1956 and 1975, though it remained important as a flexible, nonpoliticized source of funds for the high-growth sectors.[59]

But the early priority given to the basic raw materials processing industries, social overhead investments, and then to heavy machinery and assembly sectors meant that Japan's semiconductor industry benefited relatively little from preferential financing when the power of industrial policy was at its height. It was not until the mid-seventies, following the first oil shock, that microelectronics rose to the top of priorities set forth by industrial policy.[60] But by that time, capital was no longer in short supply, Japan's industrial economy had grown prodigious, and pressure for internationalization had gathered momentum. Thanks to structural maturation, therefore, Japanese industrial policy is no longer as needed or as effectual today as it once used to be.

Are the subsidies excessive? That depends on one's standard of comparison. If one compares the amounts with what governments in other countries (including the U.S.) supply, the answer is no. France and Great Britain provide larger subsidies;[61] and the U.S. federal government much larger sums.[62] Since every weapon system now in the works or planned will require embedded computers (a growing portion of which will be microprocessors), the dollars returned to U.S. industry for design, development, and production will be huge. In 1983, the estimate of dollars returned to the U.S. electronics industry came to $8 billion; by 1990, it could go as high as $33 billion.* The armed services seem to be increasingly inclined to contract out to industry rather than trying to take on computer design and development tasks on their own.[63] Of course, not all of this is of direct commercial benefit; but even applying a generous discount factor to adjust for low spill-over effects, the sums still make Japanese subsidies look small by comparison. Between 1976 and 1982, Japanese government subsidies averaged about $50 million a year, or less than 10 percent of total R&D. Estimates for national projects currently under way, or on the drawing boards, until 1990 come to a cumulative multiyear total of anywhere between one and two billion dollars, depending on the amount eventually spent on the Fifth-Generation Computer Project. It seems a bit odd, therefore, to hear complaints about excessive MITI subsidies being voiced by governments whose own spending levels exceed that of Japan's.

As for the charge that, by reducing risks, Japanese industrial policy breeds abnormal rates of investment and excess capacity, the answer is less clear. Looking back on the postwar evolution of Japan's steel and petrochemical industries, one can find confirmation of this claim. Yet aggressive capital investment is a phenomenon common to priority and nonpriority industries alike. Look at consumer electronics (color television, VTRs, hand-held calculators, etc.) and other consumer durables (cameras, watches, precision instruments, etc.) or certain chemical products: the investment competition has not been discernibly less severe than in certain priority areas (e.g., biotechnology). Is industrial policy the cause? Or is it Japan's financial system? Or is the real reason low inflation, low interest rates, and a predictable investment climate?

Is investment in priority industries really risk-free? That, too, is

---

* This estimate includes all electronic components and systems, not just semiconductors. Disaggregated data for semiconductors alone are unavailable, but they would be only a fraction of the total.

open to debate. For industries hard hit by cyclical or structural recession, the Japanese government has been known to provide some measure of downside relief by legalizing the formation of anti-recession cartels.[64] High export elasticity has also cushioned some of the risks of heavy capital investments. Therefore, to the extent that Japanese industrial policy has facilitated cartelization and export-led recovery from recession, it has helped to reduce investment risks. But anti-recession cartels have tended to be formed for mature, capital-intensive industries, suffering from excess capacity worldwide. For growing industries like semiconductors, potential government backstopping in the form either of cartelization or of loan guarantees (already discussed) cannot be considered an elimination of investment risks. Overseas markets are also no longer fully permeable for Japanese exports. And the ripple effects of indicative lending, another feature of Japanese industrial policy, are not nearly as far-reaching as before, thanks to the greater availability of capital.* Hence, it is not clear how much of the investment imperative in Japan's semiconductor industry can be linked directly to the existence of industrial policy.

The combination of heavy capital investments (whatever the cause) and national research projects, however, has led to the phenomenon described abroad as "export downpours." National research projects tend to bring participating companies to the starting lines of product commercialization around the same time. Instead of staggering the introduction of new products, companies tend to come out with similar products at the same time. The domestic market is quickly saturated, and overseas markets are then "deluged" with Japanese goods pouring out of modern, efficient, and large plant facilities which aggressive capital investments put in place.† The high fixed costs of lifetime employment and heavy debt-servicing make it difficult to allow significant levels of idle plant capacity.

The problem of "export downpours" from Japan has not been alleviated by the ad hoc, highly politicized reactions of the U.S. government. Indeed, the problem has been magnified by the sequential imposition of U.S. trade restraints on one Japanese industry after another. Each time ceilings have been clamped on one area, the Japanese

---

* Indicative lending refers to the cues that government lending to priority industries signals to private banks in Japan. Small loans from the Japan Development Bank might make city banks willing to invest more in that industry because it is perceived to be a fairly safe risk.

† Here again causal attribution is difficult. Export downpours have also come from nonpriority industries, with no history of national research projects, that produce such items as color television sets, VTRs, and zippers.

have merely taken up the slack by stepping up exports in other, higher value-added areas. The composition of exports has shifted, accordingly, from textiles, to steel and color televisions, to automobiles, and now increasingly to high technology. Over half of all Japanese exports to the U.S. now fall into product categories for which trade ceilings have been established. Yet the total influx of Japanese goods has kept going up. This suggests that America's trade restrictive stance has inadvertently helped Japan adjust its industrial structure to ever higher value-added production, toward which its comparative advantage has been shifting anyway. It has played to Japan's strengths while encouraging inefficiency at home. In this sense, America's semiconductor industry has been rendered doubly vulnerable by the combination of Japan's "torrential" exports and the perverse effects of U.S. trade restrictions in the old-line manufacturing sectors.

Mass production and greater flexibility with respect to acceptable return on investments permit Japanese companies to price aggressively, often below what foreign competitors are charging. American firms are placed under severe price pressures. Their choice is either to compete on prices and feel the squeeze on profits, or give up market share. The SIA report says the U.S. semiconductor industry has already felt the effects of the first wave of "export deluge" in its 16K RAM and 64K RAM markets. According to SIA estimates, the net losses for five U.S. producers came to nearly $150 million over only a two-year period, 1981-82.[65] It is hard to assess the accuracy of these estimates, since the 1981-82 recession may have had something to do with the losses; but there is no doubt that Japanese price competition has put the squeeze on profits in mass volume memory devices, a source of revenue as well as a technological base for competitiveness in other product areas.

Japanese firms have set 64K RAM prices so low as to raise suspicions of dumping.[66] Some American executives find it hard to believe that Japanese firms, selling at such low prices, could even break even, since the Japanese do not appear to hold major cost advantages in terms of production equipment, economies of scale, or cumulative production experience. The Japanese may have certain advantages owing to the lower costs of capital, discussed earlier, or to the manner in which diversified Japanese firms handle internal accounting for semiconductor costs. MITI officials, concerned about harmonious U.S.-Japanese relations, continually warn Japanese companies against dumping practices, harping on the fact that it would backfire and cause serious damage to everyone.[67] But in the absence of cost data, broken down into semiconductor and other divisions, it is impossible to reach definitive conclusions. U.S. companies are thus left in something of a dilemma: privately convinced that dumping is taking place but reluctant to file formal charges because of the elusiveness of evidence and the general ineffectiveness of antidumping proceedings in the U.S. government. The ineffectuality of the antidumping system limits its deterrent value and leaves domestic industry vulnerable.

How, then, do we evaluate criticisms of Japanese industrial policy? The criticisms seem to assume, or at least strongly imply, that the objectionable features associated with Japanese industrial policy are (1) accurate characterizations of current government policies, (2) direct consequences of industrial targeting, and (3) unfair in the sense that such practices are either peculiar to Japan or found there in more virulent form. All three assumptions can be called into question. Some of the charges brought against current policies—such as the imposition of tacit R&D and production cartels—simply do not reflect the realities of semiconductor competition (though the history of other industries offers some substantiation).

Especially problematic is the second assumption. Few of the criticisms demonstrate a direct link between a specified cause (e.g., targeting) and some effect (e.g., overinvestment). As pointed out earlier, not everyone agrees that industrial policy has made a decisive difference in the competitiveness of Japanese enterprise. The controversy is far from settled and awaits further empirical analysis. One way of shedding light on the controversy is to ask a hypothetical question: would Japan's semiconductor industry be able to compete if it had to do without those industrial policies that have been labeled unfair? Probably it would. Most of the policy measures that have given rise to the loudest complaints are not central: specifically, "targeted" loans, R&D subsidies, closed procurements, alleged cartelization, "Buy Japanese" propensities, "risk-free" capital investments, and so forth. Shorn of these putative advantages, Japan's semiconductor companies would still be able to compete effectively. The specific mix of policy measures is adjustable. If some are abandoned, others can be devised.

But the semiconductor industry would be badly hurt if the structure of government-business relations were to deteriorate. This is where the mechanism of Japanese industrial policy (not its specific content) serves its most constructive function. It is a linchpin in the whole structure of close government-business relations, and the enduring edifice of that relationship is what counts, as pointed out in Chapter 4.

To focus only on certain policy measures is to lose sight of the larger context within which industrial policy functions.

The third assumption—that certain policy measures are peculiar to Japan—is belied by the actions of the French state, as mentioned earlier. If there is anything unique, it is the fact that Japanese industrial policies have been largely successful—so successful, indeed, that the American government has moved to adopt similar policies. The Defense Advanced Research Project Agency (DARPA) has organized its own version of a multiyear supercomputer project in response to MITI's Fifth-Generation Project. MCC (Microelectronics and Computer Technology Corp.) is organized as a research collectivity so as to reduce R&D redundancy and pool the resources of individual companies. Other policies include closed procurements (for the military), generous tax incentives, and large R&D subsidies. Americans have adopted such measures in order to neutralize the perceived advantages that the Japanese derive from industrial policy; but the U.S. semiconductor industry may discover, as the Japanese already have, that although such policies can be helpful, they offer no assurance of competitive advantage. Nevertheless, it is interesting—perhaps ironic—that the U.S. government is, on the one hand, applying strong pressure on Japan to alter or abandon unfair features of industrial policy yet is simultaneously adopting selective measures that are similar to those being practiced in Japan. Nothing more clearly illustrates the seriousness with which the Japanese challenge in semiconductors is being taken.

The functions of industrial policy. If the need for, and power of, Japanese industrial policy has waned over time, and if it gives rise to foreign criticisms and trade conflicts, why is it maintained? What positive functions does it serve? On a practical level, industrial policy provides MITI with its raison d'être and main vehicle of influence. For that reason alone, it stands little chance of being abandoned (U.S. pressures notwithstanding). Bureaucratic power is at stake. Were MITI to lose its power to implement industrial policy, it would lose much of the legitimacy it now possesses to coordinate government-business interactions, mediate conflict, mobilize consensus, and formulate a "vision" of the structure of Japan's economy five or ten years down the road. Industrial policy is therefore essential from MITI's perspective. Chapter 4 identifies a number of other vital functions:

—protection against politicized interference in micro-industrial management

—control over special interest lobbying; industrial and national aggregation

—reduction of uncertainty and risks

—sanctioning of agreed upon industrial objectives

—mobilization of public understanding and support for industrial
objectives

—facilitation of consensus within companies

—compensation for market deficiencies

—industrial location

—indirect effect of pointing the country's best talent toward high-
growth industries and sectors

—improvement of information available

—enhancement of efficiency

—industrial restructuring in response to shifting comparative
advantage

—international conflict avoidance and mediation; crisis management

If Japan's industrial policy has facilitated the growth of its semiconductor industry for the reasons listed above—not because of the
stereotypes usually invoked such as "Japan, Inc."—has the absence of
an official industrial policy crippled the U.S. semiconductor industry?
To answer this question requires an assessment of what is missing in
the industrial policy literature, namely, a clear and systematic identification of that which industrial policy is supposed to affect—that is,
the dependent variables. There are at least seven, and probably more,
that the chapters in this book touch upon: (1)taxes, (2) corporate finances, (3) capital investments, (4) technological development, (5) industrial structure, (6) trade, and (7) political consequences. Let us
evaluate each briefly.

Taxes. The U.S. semiconductor industry is not at a disadvantage because of its corporate tax burden, even though it lacks a few special
features like the faster depreciation allowance and tax-free retirement
reserve fund that benefit Japanese industry. The U.S. semiconductor industry is hurt indirectly by the pernicious effects of inflation and by
the generous tax breaks given inefficient but politically powerful industries. Contrary to common misperception, inter-sectoral variations
in corporate taxes are much wider in the U.S. than in Japan, notwithstanding stereotypes of discriminatory "targeting" in tax policies.[68]

Corporate finances. U.S. capital markets have channeled an abundance of money to the semiconductor industry. As discussed earlier, Japan's financial system may provide easier or steadier access to capital at lower costs, with more flexibility concerning expected return on investments. But one price has been the slowness and difficulty of establishing a venture capital market. Japan's "targeted" industrial policy has had its advantages and drawbacks.

Capital investments. Both U.S. and Japanese semiconductor industries are investing heavily in R&D and new plants and equipment. U.S. industry holds a wide lead in absolute amounts invested each year, though the Japanese are investing at a faster pace. So far, at least, the lack of an explicit industrial policy has not hampered the American side.

Technological development. Technological "targeting" has helped the Japanese focus on clear goals and mobilize public and private resources to achieve them. National research projects have accelerated the catch-up process. But there is no guarantee that, because the VLSI project succeeded, subsequent efforts like the Fifth-Generation Supercomputer Project will, too. A major price Japan has paid for its industrial policy is the comparatively weak record of product innovation in its small-medium sector. The U.S. industry—which includes IBM and Texas Instruments as well as small merchants houses—is the world's technological leader. Again, America has managed very well without an industrial policy—perhaps better than if the government had extended a more visible and dominant hand.

Industrial structure. Taken as a whole, U.S. industrial structure is a source of technological ferment and commercial competitiveness, a national asset of enormous importance. The whole stratum of small-medium merchant firms, so central to the evolution of semiconductor technology, is missing in Japan. Some executives of small firms in the U.S. fear that the adoption of an extensive industrial policy would discriminate against them in favor of big producers; the fear is not without foundation.

Trade. Japanese industrial policy has always emphasized exports. Japan has developed formidable export capabilities in terms of cumulative experience, trained manpower for overseas work, and a system of trade-related institutions. U.S. semiconductor firms have done well in Europe, thanks in part to the foothold provided by direct investments there. Past investment restrictions in Japan have hampered America's efforts to establish a beachhead from which to attack Japan's large and potentially lucrative market. U.S. semiconductor firms have also been hurt by the imposition of trade-restrictive measures on Japanese imports of steel, automobiles, and other products of smoke-stack industries.

Political consequences. Industrial policy in Japan fits into, and rein-forces, a system that insulates management of the industrial economy from partisan politics. In America, economic policy is politicized and costly from the point of view of economic efficiency.

So far, except in the last two categories, America's lack of an explicit industrial policy has not hindered its semiconductor industry either in terms of its growth or its competitiveness. A strong industrial policy like Japan's could help perhaps, particularly in view of inefficiencies bred in other parts of the U.S. industrial economy; but unless institu-tional changes are made (e.g., the strengthening of institutions for in-terest aggregation), the formal adoption of an industrial policy would be apt to create more problems than it would solve.

Like other efficient industries, semiconductor manufacturers bear the costs of inefficiency caused by America's backward-leaning poli-cies. The U.S. semiconductor industry would benefit if such costs could be eliminated. Indeed, improvements in current policy would probably be more helpful than attempts to adopt an official industrial policy on the Japanese model. The semiconductor industry would also benefit if, in addition to stepping back from policies which breed in-efficiency, the U.S. government were to:

—provide more support for basic R&D
—raise rates of capital formation
—support education and training in basic sciences and engineering
—facilitate structural adjustments to changing comparative advan-tage
—upgrade the institutional infrastructure to deal with trade-related issues (e.g., more effective antidumping procedures)

Such measures, if undertaken, would offer the U.S. semiconductor in-dustry some of the same benefits enjoyed by its Japanese counter-part—without having to transplant an industrial policy that is not particularly well suited to America's political and economic system.[69]

Industrial policy has worked in Japan, and practically nowhere else, largely because of the mutually reinforcing and complementary nature of national institutions, practices, and values, from the structure of the politico-economic system down to the microlevel units of company and individual.[70] In many industrial systems, state and society, public policies and private enterprise, collective interests and corporate goals, societal dynamics and individual socialization, socio-cultural patterns of behavior and the functional needs of the capitalist system often coexist in a state of latent tension, if not overt contradiction.[71] But such internal conflicts, though to some extent unavoidable, somehow seem to have fit together with less dysfunctional friction in Japan's pursuer system. Japan's extraordinary rate of capital formation, and its distinctive mechanisms for allocation, for example, have supplied the wherewithal to sustain high levels of capital investments, which industrial policy has helped to channel. Capital investments, in turn, have led to steep increases in productivity and heightened international competitiveness. New plant facilities have been used effectively by Japan's skilled labor force, which has excelled at process technology and manufacturing. For latecomer countries, the value of production prowess can scarcely be exaggerated; it is the primary means of overtaking frontrunners.

The stability and effectiveness of Japan's political institutions also deserve special emphasis.[72] The comparative insulation of Japan's industrial policy from partisan intervention and issue-specific capturing by vested interests is one of its most striking and distinctive features, as Chapter 4 points out. It is one of the central reasons why Japanese industrial policy has worked. Japan's gerrymandered electoral districting provides a partial, structural explanation for the orderliness of interest aggregation. Politically powerful groups (e.g., farmers) are not economically dominant; economically powerful interests (e.g., banks, electronics companies) are not the most politically active or influential. Interest group power is structured in ways that defy linear product-cycle interpretations of political coalitions and outcomes.[73] Economic ministries are not simply captives of big business, dominant class interests, or the ruling LDP.[74] They act as guardians of industry-wide and public interests.

As the external environment changes, however, it is not clear how well Japan's political-economic system, heretofore geared to full-speed growth, will be able to adapt. The number of interest groups has proliferated as Japan's industrial economy has expanded, complicating the processes of aggregation.[75] Political parties have come to exercise greater influence over economic policymaking.[76] Will Japanese industrial policy be as effective for high technology industries as it has been for smokestack sectors? Will it retain its nonpoliticized nature? Can Japan overcome the weaknesses of its pursuer system as it moves toward becoming a pioneer? Will it establish a lively venture capital market and a stratum of highly innovative small-medium enterprises? If it does, what impact will this have on Japan's labor market, employment practices, corporate strategies, inter-firm competition, industrial structure, transnational linkages (since foreign venture capital is sure to flow into Japan), and world competition? Japan's industrial system is clearly caught up in an epoch-making transition from "smokestack" to high technology, from pursuer to pioneer.

Summing up. This book has analyzed the major forces at work in the U.S.-Japanese semiconductor competition, perhaps the first of its kind to arise as both countries make the transition to a knowledge-intensive economy. The semiconductor case may be a more telling barometer of what lies ahead than past bilateral experiences. For this and other reasons, it is a fascinating and complex new chapter in the postwar history of economic interactions involving the U.S. and Japan.

Two caveats are in order here. The first is that this study has not treated all facets of the semiconductor competition exhaustively. Such topics as corporate management and strategy and international trade and investments, which are absolutely central to the outcome, have been dealt with only in passing. Trade data is notoriously hard to evaluate, since semiconductor production crosses national boundaries (overseas fabrication or testing, etc.). Accurate assessment of how much value is added at each stopping point is usually not available. Corporate management is easier to investigate, and public interest in comparative management styles, particularly Japanese management, is high.[77] But like other aspects of the two semiconductor industries, management in each country is sufficiently diverse that national generalizations can only be advanced at the risk of simplification. Management—like trade, marketing, and distribution—is of crucial importance and clearly requires fuller analysis than this book has been able to provide.

The second caveat deserves even stronger emphasis. The approach taken here—national systems in competition—hopefully sheds light on the broad context within which the two semiconductor industries operate. But the national approach is artificial in the sense that it selectively focuses only on those elements that affect each side in the bilateral competition—when, in fact, the dynamics of the competition are more complicated than that. Competition *within* each country is at least as keen as that *between* countries. NEC is locked in a fierce struggle with Hitachi and Fujitsu, just as Motorola and Texas Instruments vie vigorously against each other for market share.

Because the semiconductor market is international, corporate actors and some of the commercial forces at work cannot be confined to purely national categories of analysis. In addition to familiar features of multinationalization (associated with old-line industries) such as direct foreign investments and offshore production,[78] a variety of other transactions have blurred national boundaries. One sees a proliferating network of cross-licensing arrangements, second-sourcing, joint production, technological cooperation, inter-firm transferability of certain technologies (e.g., CAD/CAM), joint use of marketing and distribution channels, OEM agreements, and joint sales based upon complementary product portfolios. If U.S. firms accept the Japanese government's open invitation to join in national research projects, one will even see joint R&D.

The opportunities and incentives for transnational cooperation are manifold—much greater than was characteristic of old-line manufacturing industries. Companies in the smokestack sectors multinationalized in order to utilize lower-cost labor, gain closer access to raw materials, circumvent import substitution policies, and take advantage of the tax and tariff benefits of moving overseas. The threat of trade restriction has been a primary factor behind bilateral cooperation in the auto industry. Without the threat of protectionism, Japanese auto makers may not have been willing to enter into joint agreements; nor, without the loss of industrial leadership and bleak business prospects, would U.S. companies have agreed.

For high technology industries like semiconductors, the inducements are more positive.[79] Concern about trade restriction is a factor, to be sure, but there are other inducements of more decisive significance, such as forward and backward integration, product differentiation, technology bartering, cross-licensing, greater familiarity with foreign customer needs, information about foreign competitors, recruitment of skilled manpower abroad, access to favorable foreign financing, economies of scale in distribution, and effective servicing. So compelling are the inducements, indeed, that the shift to an economy based on high technology could bring a mushrooming of transnational linkages.[80]

Certain contingencies like a severe contraction of trade could of course constrain international cooperation. The extension of transnational ties ("transnationalization") is contingent on vigorous competition and relative openness in the world system—though the threat of closure can set off a flurry of foreign investments and other measures of circumvention. If the world system stays relatively open, recent trends indicate that the high technology sector (e.g., the aircraft and computer industries) will develop extensive patterns of crisscrossing linkages. If so, the weight of national factors will be commensurately reduced. And even framing the semiconductor competition in terms of a U.S.-Japan confrontation will be somewhat artificial. "Transnationalization" has not proceeded that far yet, but this caveat must be borne in mind as the reader digests the implications of the research findings presented in this book.

We have found that the world's two largest semiconductor producers, the U.S. and Japanese industries, possess strengths in virtually all areas of competition. This study suggests that each demonstrates special capabilities closely related to its industrial system. U.S. industry is especially strong in the area of product innovation, venture capital markets, industrial structure, technology diffusion, the cumulative experience and know-how of individual firms, world market share, university-based research, and the world dominance of related downstream industries, particularly computers and telecommunications. Its comparative advantages lie, in short, in the semiconductor industry itself and the traditional dynamism of America's high technology sector. Japan's strengths reside in its political-economic system: its resilient economy, high rate of capital formation, sound macroeconomic performance, political system, nonpoliticized management of its industrial economy, commercial orientation, and banking-business-bureaucracy infrastructure.

The findings of this study paint a picture of industrial competition that is far more complex than the usual black-white portrait. Take technology for example. This study has found that contrary to prevailing stereotypes, the Japanese seem to be making progress in their capacity to innovate in state-of-the-art technology, just as Americans have shown that they can compete in quality. Technological innovation, we believe, is best understood as a function more of "structural" and "market" forces than of "cultural" traits or "unique" factors. Hence, the most revealing indicators of future innovation are trends in market demand, technology push, venture capital, industrial and corporate organization, and strategy, not diffuse cultural characteristics like diligence or Japanese labor market practices such as lifetime employment and the seniority system. Perhaps Japan's most serious deficiency in terms of its institutional structure for innovation is its lack of a venture capital market. If it is able to establish a venture-based system, with all that this implies for entrepreneurship and industrial structure, the effects on Japan's innovative capacities (as well as transnational linkages) could be far-reaching.

This study has also found evidence that strongly qualifies, if not directly contradicts, stereotypical notions of corporate behavior, finances, MITI's role, and industrial policy. Although U.S. semiconductor firms pay closer attention than their Japanese counterparts to quarterly profits, they are not simply maximizers of short-term profit but realize that high levels of capital investment in both R&D and equipment must be sustained in order to stay competitive. They are well aware of the importance of world market share, because the semiconductor industry happens to be one in which the learning curve effect is quite pronounced. Unlike U.S. auto and steel manufacturers in the past, semiconductor producers realize that they must penetrate the Japanese market early and not allow Japanese producers to monopolize their own markets, otherwise Japanese firms will use economies of scale and cumulative production experience to launch an assault on the U.S. (and third) markets.

Although it is true that Japanese firms put market share at the top of their objectives—with capital gains for stockholders at the bottom—they are by no means indifferent to earning profits.[81] Nor can they afford to be, given the fact that Japanese banks monitor trends in profit earnings more carefully than is widely believed. Japanese companies are probably under less compulsion to maximize profits, but they do try to show some level of profitability, even if, in the eyes of American executives, the threshold often seems to be set fairly low.

Owing to a larger portion of external indirect financing, Japanese companies are also not subjected to as much pressure on ROI as American firms. But ROI ranks near the top in the hierarchy of priorities established by most Japanese companies, and the widespread assumption that lower ROI pressures permit Japanese companies to pursue a longer term strategy while U.S. firms are doomed to chase the phantom of short-term profit is a simplification. While U.S. firms are more concerned about achieving high ROIs, this does not necessarily preclude them from following long-range plans to expand market share. Indeed, because of the learning curve phenomenon, most semiconductor companies—American and Japanese alike—must factor market share into their strategic calculus, or risk being forced out of the competition by those that do.

The financial picture is also more complicated than stereotypes usually convey. It is true that Japanese firms have probably had access to lower-cost capital over the past several years, though the comparative costs of capital are notoriously hard to quantify accurately. The reason has had less to do with the structure of debt vs. equity financing than with Japan's higher rate of savings and its macroeconomic policies. How much of an advantage Japanese firms enjoy is hard to say, because capital costs change constantly, and the burdens of heavy dependence on bank loans are not normally brought into the equation. The desire to reduce their vulnerability to interest rate fluctuations (under a more internationalized financial system) explains what is otherwise baffling: namely, the trend away from low-cost deficit financing in Japan. In spite of America's low rate of capital formation, the U.S. semiconductor industry has not been strapped by a lack of capital. America's capital markets have functioned fairly well in meeting the spiraling needs of the semiconductor industry. Indeed, capital investments in the U.S. semiconductor industry exceed those in Japan by a wide margin. Where Japanese firms have an edge is in the country's rate of capital formation and in the flexibility associated with its banking-business infrastructure.

As for the government's role, the attributes most commonly cited in connection with Japan's postwar economic performance—"Japan, Inc.," MITI, and industrial policy—explain only part of the story (and largely an early part at that) behind the successful growth of Japan's semiconductor industry. The role of MITI in particular tends to be glorified, perhaps because of the power MITI wielded during the fifties and sixties. MITI does not have the power today (analogous to that of a giant monopoly) to impose a collusive division of labor on the semiconductor industry. Its main contribution lies in the creation of a buoyant business environment that offsets market deficiencies and gives semiconductor firms positive incentives to compete in the international marketplace. MITI will undoubtedly do what it can to aid the future development of the semiconductor industry, but the principal dynamism behind the industry's growth will continue to come from the private sector.

So far, at least, the U.S. semiconductor industry has not been crippled by the lack of an official industrial policy. It has been able to maintain its position of world leadership. At the same time, however, Japan's rapid emergence may mean that the mere continuation of past policies is not enough to hold off the challenge. New policies, new institutions, and better coordination may be called for, particularly in the areas of trade and control over politicized intervention. But such changes need not assume the form of an explicit industrial policy. Correcting some of the most egregious inefficiencies in America's de facto industrial policy may be preferable. Probably the most effective treatment of all would be the strengthening of macroeconomic management.

The competitive challenge posed by Japan's rapid growth in semiconductors and other areas of high technology is, on the whole, a healthy development—in spite of the hand-wringing, charges of "unfairness," and occasional prophecies of doom heard in the U.S.* Stepped-up competition caused by Japan's emergence has had the salutary effect of quickening the pace of technological progress, lowering costs, and heightening efficiency. Not only have end-user companies everywhere benefited from breakthroughs in product technology, lower costs, and higher reliability; but the consumer publics in the U.S. and Japan have been the prime beneficiaries. The range of consumer products available has expanded; the material quality of life has improved; industrial productivity and economic growth have been enhanced; military capabilities have been upgraded. At the same time, of course, a variety of problems have also arisen: possible corporate bankruptcies, pressures especially on U.S. small-medium merchant houses, loss of employment, technology transfers to Communist-bloc countries, trade conflicts, and such. Potential problems should not be minimized, but they are, we think, outweighed by the benefits that have already accrued and that intensified competition promises to yield. Especially encouraging is the fact that, unlike the steel, auto, and color television industries, demand for semiconductor products is still rising rapidly, and for the foreseeable future, gives no indication of leveling off. This means that there is room for both sides to prosper. The opportunity is open, therefore, for the two semiconductor industries to chart new directions in what, with the intensification of competition in high technology, is clearly a new era in the postwar history of U.S.-Japanese relations.

---

* To the extent that charges of unfairness are valid, of course, tough measures need to be taken to see that they are corrected.

Appendixes

Kinds of Semiconductor Devices

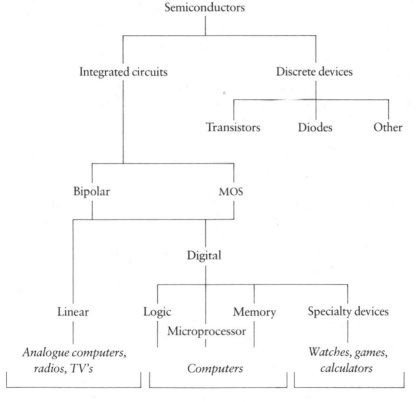

SOURCE: Robert W. Wilson, Peter K. Ashton, and Thomas P. Egan, *Innovation, Competition, and Government Policy in the Semiconductor Industry* (Lexington, Mass.: D. C. Heath, 1980), p. 20.

University of Tokyo Undergraduate Curriculum in Electrical Engineering, 1981

The three major fields covered in the curriculum of the undergraduate program in electrical engineering at the University of Tokyo are computer and communications engineering, material and device engineering, and power and control engineering. About half the students are oriented toward computer hardware and software. On the whole, the Japanese curriculum is much like that of a basic American curriculum. New courses can sometimes be introduced more rapidly into a U.S. curriculum: at M.I.T., for example, there is a very rapid build-up and test of new courses, with continual modification. But most universities make only slight changes from year to year in the undergraduate program, leaving the flexibility to the graduate program.

The most distinctive feature of the Japanese curriculum in electrical engineering is the requirement that undergraduate students spend the entire day in the classroom or laboratory. Students are expected to take all the courses offered. One afternoon is free, but it is customarily used for a visit to an industrial facility as arranged by the department. Because the course load is so heavy and graduate students are not available to work as teaching assistants (as they are in the United States), there are usually no outside assignments. Students are, however, required to submit reports on their laboratory work, which gives them experience in technical writing and forces them to formulate their own opinions. It has been observed that the extraordinary amount of time that students spend with their classmates tends to promote group cohesion and loyalty.

TABLE B.I

Core Curriculum for Electrical Engineering

Second year, second semester	Third year, first semester	Third year, second semester	Fourth year, first semester
General View on Electrical Power Engineering		→Electric Power Generation Engineering	
			→Power Systems Engineering
		Electric Power Transmission and Distribution	
		Ionized Gases	→High Voltage Engineering
	Fundamentals of Electrical Machinery and Apparatus[a]	→Electrical Machinery and Apparatus 1	→Electrical Machinery and Apparatus 2
			→Fundamentals of Electrical Materials
			→Electrical Power Applications
		Control Engineering 1	Control Engineering 2

[a]Compulsory only for power electrical engineering students.

TABLE B.2

Curriculum for Material and Device Engineering and Information and Computer Engineering

Second year, second semester	Third year, first semester	Third year, second semester	Fourth year, first semester
Introduction to Quantum and Statistical Mechanics	Fundamentals of Solid State Physics	Solid State Physics 1	Solid State Physics 2
			Physics and Chemistry of Electronic Materials
	Fundamentals of Electron Devices	Electron Devices 1	Electron Devices 2
			Electron Devices 3
			Optoelectronic Devices
			Bio-Electronic Engineering
	Fundamentals of Communication and Information Engineering	Information Theory	Applied Communication Engineering
		Transmission and Signal Processing	Communication Systems and Networks
		Introduction to Systems Engineering	Communication Switching Engineering
		Mathematical Programming for Systems Engineering	

TABLE B.3

Curriculum for Power Engineering

Second year, second semester	Third year, first semester	Third year, second semester	Fourth year, first semester[a]
Mathematics 1			
Exercise in Mathematics and Mechanics	→Mathematics 2	→Mathematics 3	
	Outlines of Computer Programming and Numerical Computation	→Electronic Computers 1─┐	┌─→Electronic Computers
	Logic Circuits		Information System Engineering
Electrical Measurements 1[b]	→Electrical Measurements 2[b]		→Microwave Engineering
Electricity and Magnetism[b]		→Fundamentals of Electromagnetic Waves	→Wave Optics and Optoelectronics
		Applications of Electrostatic and Electromagnetic Fields	→Fundamentals and Application of Plasma
Theory of Electric Circuit 1[a]	→Theory of Electric Circuit 2[b]		→Theory of Electric Circuit 3
	Theory of Electronic Circuit 1	→Theory of Electronic Circuit 2	

[a]In the second semester of the fourth year, students complete the graduation thesis for their Bachelor's Degree.
[b]Compulsory for all E.E. students.

243

TABLE B.4

Courses in the First Semester of the Third Year

Day	8:30-10:00	10:15-11:45	13:30-16:00	18:00
Mon.	Fundamentals of Electrical Machinery and Apparatus	Theory of Electric Circuit 2	Fundamental Electrical Engineering Laboratory	
Tue.	Theory of Electronic Circuit 1	Logic Circuits	Fundamental Electrical Engineering Laboratory	
Wed.	Outlines of Computer Programming and Numerical Computation	Mathematics 2		
Thur.	Fundamentals of Electron Devices	Electrical Measurements 2	Fundamental Electrical Engineering Laboratory	
Fri.	Fundamentals of Communication and Information Engineering	Fundamentals of Solid State Physics	Mathematics 2	Patent Regulation

244

Courses in the Second Semester of the Third Year

Day	8:30-10:00	10:15-11:45	13:30-16:00	18:00
Mon.	Mathematical Programming for Systems Engineering	Theory of Electronic Circuit 2	Electrical/Electronic Engineering Laboratory 1	
	Solid State Physics 1			
Tue.	Electronic Computers 1	Applications of Electrostatic and Electromagnetic Fields	Electrical/Electronic Engineering Laboratory 1	
		Introduction to Systems Engineering		
Wed.	Control Engineering 1	Mathematics 3		
Thur.	Electric Power Transmission and Distribution	Electric Power Generation Engineering	Electrical/Electronic Engineering Laboratory 1	Industrial Economy
	Transmission and Signal Processing	Electron Devices 1		
Fri.	Electrical Machinery and Apparatus 1	Ionized Gases	Design in Electrical and Electronic Engineering	
	Fundamentals of Electromagnetic Waves	Information Theory		

Major Computer and Semiconductor Facilities at the University of Tokyo, 1981

Although LSI chips cannot be fabricated, graduate students at the University of Tokyo can fabricate discrete devices and small-scale integrated circuits by themselves. LSI design is not taught in any course, nor is there anything so far at any Japanese university to compare with the courses on LSI design offered at some U.S. universities. Japanese universities are seeking to introduce such courses, however.

TABLE C.1
Facilities at Computer Centers

Function	At main computer center	At educational computer center
CPU	M 200H (Hitachi) × 8	MELCOM 900 (Mitsu-bishi) × 1 (MELCOM 900 II × 1 Sept. 1981)
Main memory	16M Byte/pair	1.5M Byte × 1 (6M Byte × 1 Sept. 1981)
Main memory device	MOS 16K bit	MOS 4K bit
Disc memory	317M Byte × 96 200M Byte × 32	200M Byte × 6
Mass storage system	34.8G Byte	
Processing rate	8 ~ 12 MIPS × 8 (~ 24 MIPS × 6 with Array Processor) 4 MFLOPS × 8 (9 MFLOPS × 6)	1 MIPS

TABLE C.2

Major Capabilities of the Semiconductor Processing Laboratory

Thermal oxidation	Photo-lithography
Thermal diffusion	Auger electron spectroscopy
Ion-implantation	Electron spin resonance
Reactive ion etching	Xray diffraction
Plasma assisted vapor deposition	Electron beam diffraction and
Plasma anodization	microscopy
Electron beam evaporation	Clean room (200 m²)
High frequency sputtering	Electronic and optical measurement

Glossary of Abbreviations
and Technical Terms

The definitions of terms are based on Robert Wilson et al., *Innovation, Competition, and Government Policy in the Semiconductor Industry* (Lexington, Mass.: D. C. Heath, 1980), and Mel H. Eklund and William I. Strauss, eds., *Status '80: A Report on the Integrated Circuits Industry* (Scottsdale, Ariz.: Integrated Circuit Engineering Corporation, 1980).

AEA. American Electronics Association

AMD. Advanced Micro Devices

AMI. American Microsystems, Inc.

AT&T. American Telephone and Telegraph

bipolar. One of two main types of transistors (along with MOS). Bipolar transistors were the dominant semiconductor devices of the 1950's. They operate at higher speeds than MOS devices, making them especially useful for such signal processing as radar and communications

CAD. Computer-aided design

CAM. Computer-aided manufacturing

captive market or supply. Production exclusively for in-house consumption; usually contrasted to merchant market or supply, which involves the sale of semiconductors to other companies

CDL. Computer Development Laboratories comprised jointly of Hitachi, Fujitsu, Mitsubishi as part of Japan's VLSI Project

CMOS. Complementary Metal Oxide Semiconductor; possesses n-channel (negative-conducting properties) MOS transistors and p-channel MOS transistors on the same chip; known for low power dissipation and density of elements per unit area

cross-licensing. An agreement between two (or more) companies to exchange technologies instead of unilaterally selling patent rights or technological know-how at contracted royalty rates

custom circuits. Integrated circuits specially designed to individual customer specifications, as opposed to standard-design integrated circuits, which are produced in volume for general use in a variety of equipment designs; in the late 1960's, it was thought that the future of digital integrated circuits lay in custom LSI designs because product function and performance could be differentiated through custom designed circuits. For a variety of reasons, including the invention of the microprocessor, this prediction has proved wrong.

DARPA. Defense Advanced Research Projects Agency, Department of Defense

digital ICs. One of two major types of integrated circuits (the other being linear); operates on on-off switching properties of transistors and diodes to store data and perform logical operations using binary mathematics; digital technology is the basis for such advanced ICs as logic and memory chips and microprocessors, which are installed in computers.

DOD. Department of Defense

EIAJ. Electronics Industry Association of Japan (*Nihon Denshi Kikai Kogyokai*)

electron beam processing. The use of electron beams to expose photosensitive lacquers in the manufacturing of masks for very fine line (measured in microns) photolithography and also for processing photosensitive materials directly on the semiconductor wafer avoiding separate masks.

FILP. Fiscal Investment and Loan Program (*zaisei toyushi*)

FTC. Fair Trade Commission (*Kosei Torihiki Iinkai*)

gallium arsenide. A compound semiconductor material. Its prime virtue, and the reason it is receiving significant research attention, is very fast speed (much faster than silicon). Gallium arsenide is potentially very attractive as an alternative to silicon-based semiconductor technology, not unlike that of Josephson junction, another very high speed device.

GATT. General Agreement on Tariffs and Trade

germanium. A chemical element like silicon, used in transistors during an early period of semiconductor development; displaced by batch-produced silicon transistors

GME. General Microelectronics

IBM. International Business Machines

IEEE. Institute of Electrical and Electronics Engineers

integrated circuits (ICs). An array of transistors and diodes on a single piece of silicon crystal that is interconnected in such a way as to allow it to perform the function of a complete electronic circuit. ICs are the largest and most important cluster of devices in the semiconductor category.

ion implantation. High-voltage bombardment of ions on semiconductor wafers in precise patterns producing the semiconductor structure appropriate to the desired electronic device.

ITC. International Trade Commission

JDB. Japan Development Bank (*Nihon Kaihatsu Ginko*)

JECC. Japan Electronic Computer Corporation (*Nihon Denshi Keisanki Kabushiki Kaisha*)

JEIDA. Japan Electronics Industry Development Association (*Nihon Denshi Kogyo Shinko Kyokai*)

Josepheson junction. A potential superconductor that could supplant semiconductors. It has the possibility of switching speeds twice as fast as the fastest semiconductors while using only one-thousandth the energy, and also of being packed more densely. So far, most of the experimental devices have used lead alloys, especially lead-indium-gold, as the superconducting elements in the circuits. The drawback to Josephson junction electronics is that the superconductors must be cooled to very low temperatures, near that of liquid helium, and the cycling between cryogenic and room temperatures creates great stresses that can damage the lead alloys; niobium or some other refractory material has a higher melting temperature and is therefore more resistant to deformation. Many problems must be worked out before the switch from semiconductors to this much faster superconductor can be made.

keiretsu. Group of affiliated companies in Japan, such as Sumitomo or Mitsubishi, centered on a major bank (e.g. Sumitomo or Mitsubishi Bank) or a major manufacturing enterprise (e.g. Toyota); many semiconductor producers in Japan belong to a *keiretsu* (e.g. NEC belongs to the Sumitomo group).

LDP. Liberal Democratic Party (*Jiminto*)

learning curve. The lowering of production costs with increases in output volume. For certain mass-produced memory chips, production costs have tended to fall by around 30 percent with every doubling of output. This phenomenon accentuates the importance of early entry in new product markets and holds out incentives for companies to follow a strategy aimed at expanding market shares.

linear ICs. Unlike digital ICs, which function in terms of two states (on and off), linear ICs operate on the basis of continuous signals; linear devices are used in a variety of consumer electronics products such as radios, television sets, and audio equipment.

logic devices. Semiconductor chips (digital integrated circuits) that perform arithmetic functions or make certain decisions; such functions are made possible through the interconnection of components to form logic gates in a variety of circuit designs.

LSI. Large-scale integrated circuits

MCC. Microelectronics and Computer Technology Corporation

memory devices. Integrated circuits used to store binary data; memory devices are classified by information storage capacity (64K, 256K, etc.) and by accessibility (serial or random access).

merchant house. A company that makes semiconductors and sells them to other firms for installation in their end-user products; to be distinguished from captive suppliers, which consume all their semiconductor production for their own end-products.

micron. A micrometer, or one-millionth of a meter; spaces between circuits densely packed on memory chips are measured in micron and even increasingly in submicron units.

microprocessor. A single IC chip that performs all the central processing-unit functions of a computer; if this chip is combined with memory and input-output ICs, it becomes a microcomputer.

MITI. Ministry of International Trade and Industry (*Tsusansho*)

MOS. Metal oxide semiconductor. One of two main types of transistors (along with bipolar); consists of semiconductor body (silicon) with silicon-dioxide gate dielectric and metal gate. MOS technology has been at the leading edge of semiconductor developments since the 1970's; its power consumption is relatively low, but it is slow.

NASA. National Aeronautics and Space Administration

NEC. Nippon Electric Company (*Nippon Denki*)

NTIS. NEC-Toshiba Information Systems, one of the research laboratory groups in Japan's VLSI Project

NTT. Nippon Telegraph and Telephone (*Nihon Denden Kosha*)

OEIC. Optoelectronic ICs

OEM. Original equipment manufacturer

RAM. Random access memory device. The term random refers to the capacity to store or retrieve data from any location in the memory chip; to be distinguished from serial access, which permits storage and retrieval only in some specific, sequential order. The storage capacity of RAMs has increased geometrically from 1K to 4K to 16K, 64K, 256K, 1M (megabit), and even 4M.

ROE. Return on equity, or the amount earned in terms of equity capital on hand

ROI. Return on investment, or the amount earned in terms of money invested

second sourcing. An agreement whereby one company is granted permission to produce something (e.g., a microprocessor) that another company has invented and brought out first. The "second source" gains access to new products and technology while the innovator gains royalties and buyer confidence in reliable delivery and service.

semiconductors. Technically, a material that possesses properties intermediate between a conductor and an insulator, such as silicon and germanium. The term encompasses a range of products, including discrete devices (e.g. diodes) and many different types of integrated circuits.

SIA. Semiconductor Industry Association

silicon-on-sapphire. A process that uses a sapphire substrate upon which a silicon film is epitaxially deposited and then etched into individual devices. Compared with conventional MOS devices, it offers a combination of low power consumption and relatively high speeds (at higher prices).

SRC. Semiconductor Research Cooperative

TI. Texas Instruments

USTR. United States Trade Representative

value added. The difference between the market price of a product and the cost of materials to make that product

venture capital. Money invested in new start-up companies prior to their going public and issuing stock.

VHSIC. Very high speed integrated circuits; also, the program by that name established by the Department of Defense

VLSI. Very large scale integrated circuits

wafer. Semiconducting material, usually silicon, on which individual chips or slices can be fabricated and cut into prescribed pieces for individual integrated circuits.

Notes

Chapter One

1. Masanori Moritani, *Japanese Technology* (Tokyo: Simul Press, 1982), pp. 159-73.
2. For a similar point of view see "Tomorrow's Leaders," *The Economist*, June 19, 1982, pp. 21-22; see also Edward A. Feigenbaum, and Pamela Mc-Corduck, *The Fifth Generation* (Menlo Park, Calif.: Addison-Wesley, 1983), pp. 136-48.
3. Richard W. Anderson, "The Japanese Success Formula: Quality Equals the Competitive Edge," in *Quality Control: Japan's Key to High Productivity*, verbatim record of a seminar held in Washington, D.C., Mar. 25, 1980, under the sponsorship of the Electronic Industries Association of Japan, pp. 18-19.
4. James Abegglen, *Management and Worker* (Tokyo: Sophia University and Kodansha International, 1973), pp. 34-35.
5. Interview with American semiconductor industry executive, January 1983, Cupertino, California.

Chapter Two

1. See "Chip Wars: The Japanese Threat," *Business Week*, May 23, 1983, p. 85; "The Race to Build a 'Super' Computer," *San Jose Mercury News*, Apr. 12, 1983; and "'Reasoning' Computers Are Next," *San Jose Mercury News*, May 17, 1983.
2. *San Jose Mercury News*, Apr. 12, 1983.
3. *Peninsula Times Tribune* (Palo Alto), Dec. 17, 1981.
4. "Chip Wars," p. 84.
5. *San Jose Mercury News*, May 17, 1983; "Chip Wars," p. 85.
6. John G. Posa, "How Japan's Chip Makers Line Up to Compete," *Electronics*, June 2, 1981, p. 114.
7. See *Byte*, May 1983, p. 496, and K. H. Kim, "A Look at Japan's Development of Software Engineering Technology," *Computer*, May 1983, pp. 26-37.
8. *Japan Economic Journal*, June 2, 1981, pp. 1, 9; *Electronic Engineering Times*, June 22, 1981, p. 4.
9. Posa, pp. 114, 118.
10. "Where Japan Has a Research Edge," *Business Week*, Mar. 14, 1983, p. 116.

11. "A Fifth Generation: Computers That Think," *Business Week*, Dec. 14, 1981, pp. 94-96. See also "Where Japan Has a Research Edge," p. 116.

12. See the remarks of Professor Edward Feigenbaum of Stanford University at the May 1983 panel on "Fifth-Generation Computers" at the National Computer Conference, as reported in *Computerworld*, May 30, 1983, p. 28.

13. *Japan Economic Journal*, June 2, 1981.

14. *Peninsula Times Tribune*, Sept. 18, 1981.

15. *San Francisco Chronicle*, Sept. 18, 1981.

Chapter Three

1. See *Asian Wall Street Journal*, Sept. 2, 1981.

2. See Gene Bylinsky, "The Japanese Chip Challenge," *Fortune*, Mar. 23, 1981, p. 116; and "Chip Wars: The Japanese Threat," *Business Week*, May 23, 1983, p. 82.

3. *Ibid.*, p. 120.

4. Gene Bylinsky, "Japan's Ominous Chip Victory," *Fortune*, Dec. 14, 1981, p. 52.

5. *Asian Wall Street Journal*, Feb. 8, 1982, and *Peninsula Times Tribune*, Mar. 15, 1982. The Semiconductor Industry Association decided not to file charges, partly because of the difficulty of proving the allegations. Since normal "learning curve pricing" shows a decline in prices by industry leaders as new technologies develop, it would be hard to prove that the Japanese were illegally forcing prices down. *Peninsula Times Tribune*, Apr. 5, 1982.

6. See *San Francisco Chronicle*, Apr. 8 and 9, 1982, and *Peninsula Times Tribune*, May 24, 1982.

7. "Two Chip Making Giants Gear Up for Recovery," *Business Week*, May 31, 1982, p. 92.

8. "Chip Wars," *Business Week*, May 23, 1983, p. 87.

9. Bylinsky, "Japan's Ominous Chip Victory," p. 55.

10. Bylinsky, "The Japanese Chip Challenge," p. 120.

11. Bylinsky, "Japan's Ominous Chip Victory," p. 55.

12. Electronics industry analyst Vincent Glinski of Drexel Burnham Lambert, quoted in *Newsweek*, Mar. 8, 1982, p. 80.

13. "New Generation of Chips Comes into Focus," *San Jose Mercury*, June 6, 1983.

14. Bylinsky, "Japan's Ominous Chip Victory," p. 57.

15. "Chip Wars," *Business Week*, May 23, 1983, p. 83.

16. James Magid, an analyst with L. F. Rothshild, Unterberg, Towbin in New York, quoted in *San Francisco Sunday Examiner and Chronicle*, May 23, 1982.

17. See *Newsweek*, Mar. 8, 1982, p. 80, and the interview with NEC's Keisuke Yawata in *San Jose Mercury*, Apr. 19, 1982.

18. See John G. Posa, "How Japan's Chip Makers Line Up to Compete," *Electronics*, June 2, 1981, pp. 115-17; Bylinsky, "Japan's Ominous Chip Victory," p. 57; *Asian Wall Street Journal*, Jan. 14, 1982.

19. *Asahi Evening News*, June 29, 1983.

20. *San Jose Mercury*, May 18, 1983.

21. *Business Week*, July 5, 1982, p. 52; *San Jose Mercury*, May 18, 1983.

22. *San Jose Mercury*, May 18, 1983.

23. "New Generation of Computer Chips Comes into Focus," *San Jose Mercury*, June 6, 1983.

24. Posa, "How Japan's Chip Makers Line Up to Compete," p. 120.

25. This discussion is based on interviews with U.S. and Japanese experts, and on Posa, "How Japan's Chip Makers Line Up to Compete," which contains a detailed technical description of work being undertaken by Japanese companies and research centers.

26. Economist Intelligence Unit Special Report no. 67, *Chips in the 1980s*, p. 11.

27. "The Boom in Tailor-Made Chips," *Fortune*, Mar. 9, 1981, pp. 123-26.

28. Jerry Werner, "NIH of a Different Sort," and "Computer-Aided Design and Design Automation for ICs in Japan," *VLSI Design* (May/June 1982), pp. 10 and 14-28; and telephone conversation with Mr. Werner. Another analyst who feels the Japanese are stronger than commonly believed in both software and microprocessors is John Shea, whose consulting firm publishes a monthly newsletter tracking Japanese technology. See the interview with Shea in *San Jose Mercury*, May 10, 1982.

29. See K. H. Kim, "A Look at Japan's Development of Software Engineering Technology," *Computer*, May 1983, pp. 26-37, for a survey of Japanese efforts. Kim suggested that Japan's software engineering technology might rival that of the United States and Europe by the end of the 1980's.

30. See Gene Gregory, "Hard Facts about Japanese Software," *Far Eastern Economic Review*, Dec. 3, 1982, pp. 48-54.

31. *Byte*, May 1983, p. 496.

32. *Ibid.*

33. "Pushing for Leadership in the World Market," *Business Week*, Dec. 14, 1981, p. 64.

34. John Shea (president of Technology Analysis Group, Inc.), cited in "Where Japan Has a Research Edge," *Business Week*, Mar. 14, 1983, p. 116.

35. J. M. Juran, "The Japanese Revolution in Product Quality," in *Quality Control: Japan's Key to High Productivity*, verbatim record of a seminar held in Washington, D.C., Mar. 25, 1980, under the sponsorship of the Electronic Industries Association of Japan, p. 8.

36. Richard W. Anderson, "The Japanese Success Formula: Quality Equals the Competitive Edge," *ibid.*, pp. 18-19.

37. Arthur L. Robinson, "Perilous Times for U.S. Microcircuit Makers," *Science*, May 9, 1980, p. 585.

38. *Electronic Engineering Times*, Mar. 2, 1981, and telephone conversation with Roger Dunn, Sept. 2, 1981, in which he corrected the figures given in that article.

39. *Electronic Engineering Times*, Nov. 10, 1980.

40. A survey by the Semiconductor Industry Association of 16K RAM prices in Japan in late 1978 found them to be more than double the prices the Japanese companies charged in the United States. See U.S. International Trade Commission, *Competitive Factors Influencing World Trade in Integrated Circuits* (Washington, D.C.: GPO, November 1979), p. 70.

41. See the comments of Dick Eichenseer, a member of Hewlett-Packard's corporate component engineering staff for digital ICs and memories, reported in *Electronic Engineering Times*, Mar. 2, 1981.

42. *Peninsula Times Tribune*, July 2, 1981.

43. See Michael Kirst, "Japanese Education: The Key to a Continuing Economic Miracle," *San Jose Mercury*, Feb. 15, 1981, and "Computing Japan's Hidden Flaw," *Cal Today*, September pub. 13, 1981, p. 9.

44. *Electronic News*, July 13, 1981.

45. *Electronic Engineering Times*, Mar. 2, 1981, and telephone conversation with Roger Dunn, Sept. 2, 1981.

46. Bylinsky, "The Japanese Chip Challenge," p. 118.

47. According to data supplied by Applied Materials, Inc., U.S. firms in 1982 spent about 15 percent of sales on capital expenditures, compared with about 23 percent for Japanese firms.

48. *San Jose Mercury*, June 6, 1983.

49. These statistics were provided by Howard Bogert of Dataquest.

50. See *San Jose Mercury*, Mar. 11, 1982, and *San Francisco Sunday Examiner and Chronicle*, Dec. 27, 1981.

51. See articles in *San Jose Mercury*, Jan. 25 and Mar. 11, 1982, quoting Bert Moyer, vice-president of finance for National Semiconductor, and John L. Lazlo, Jr., of Hambrecht and Quist, San Francisco.

52. Quoted in *Newsweek*, Mar. 8, 1982, p. 80.

53. Handel Jones of Gnostic Concepts, cited in Bylinsky, "The Japanese Chip Challenge," p. 122.

54. "Intel May Soon Compete with Its Customers," *Business Week*, Mar. 22, 1982, pp. 63-64.

55. Bylinsky, "The Japanese Chip Challenge," p. 122.

56. "Pushing for Leadership in the World Market," *Business Week*, Dec. 14, 1981, p. 84; see also "The Boom in Tailor-Made Chips," *Fortune*, Mar. 9, 1981, p. 122.

57. "The Technology That Will Create Tomorrow's Superchips," *Business Week*, May 23, 1983, p. 93.

58. Glenn E. Penister, quoted *ibid.*, p. 92.

59. Posa, "How Japan's Chip Makers Line Up to Compete," pp. 118-20, and *San Francisco Chronicle*, Mar. 3, 1981.

60. For details see "U.S. Vendors Forging Joint Japanese Ventures," *Computerworld*, May 2, 1983, p. 72.

61. *Asian Wall Street Journal*, Feb. 18, 1982.

62. *Japan Economic Journal*, June 2, 1981; *Peninsula Times Tribune*, Mar. 18 and Apr. 20, 1982.

Chapter Four

1. Herbert S. Kleinman, *The Integrated Circuit: A Case Study of Product Innovation in the Electronics Industry*. George Washington University, D.B.A. dissertation (Ann Arbor, Mich.: University Microfilm, Inc., 1966), pp. 32-132, 172-215.

2. See Robert W. Wilson, Peter K. Ashton, and Thomas P. Egan, *Innovation, Competition, and Government Policy in the Semiconductor Industry* (Lexington, Mass.: Lexington Books, 1980), pp. 151-55; and John E. Tilton, *International Diffusion of Technology: The Case of Semiconductors* (Washington, D.C.: Brookings Institution, 1971), pp. 92-97.

3. This section draws extensively on Wilson et al., *Innovation*, pp. 151-55.

4. Tilton, pp. 92-95.

5. For more on this question of inefficiency see Frederic M. Scherer, *The Weapons Acquisition Process: Economic Incentives* (Boston: Division of Research, Graduate School of Business Administration, Harvard University, 1964), pp. 68-403.

6. Tilton, p. 57.

7. Wilson, p. 155.

8. Robert Gilpin, *Technology, Economic Growth, and International Competitiveness*, prepared for the Subcommittee on Economic Growth of the Joint Economic Committee, Congress of the United States (Washington, D.C.: U.S. Government Printing Office, 1975), pp. 65-67.

9. Alexander Gerschenkron was among the first to formulate the "latecomer's" hypothesis. See Alexander Gerschenkron, *Economic Backwardness in Historical Perspective* (Cambridge, Mass.: Harvard University Press, 1962), pp. 5-51.

10. See Electronic Panel, Committee on Technology and International Economic and Trade Issues, National Academy of Engineering, draft report, "The Competitive Status of the U.S. Electronic Industry" (Washington, D.C., 1982).

11. Gilpin, pp. 70-72.

12. Seymour Melman, "Who Decides Technology?" in S. Melman, ed., *The War Economy of the United States* (New York: St. Martin's Press, 1971), p. 150.

13. "Japan's Strategy for the '80s," *Business Week*, Dec. 14, 1981, p. 53.

14. "Japanese-U.S. Balance of Trade in Electronic Equipment and Semiconductor," *Dataquest Research Newsletter*, Feb. 1983, p. 5.

15. Interviews with American semiconductor executives, July 1982.

16. For a discussion of the U.S.-Japan trade problem see Daniel I. Okimoto, ed., *Japan's Economy: Coping with Change in the International Environment* (Boulder, Colo.: Westview Press, 1982), especially chapters by Hugh Patrick, Ryutaro Komiya, and Gary Saxonhouse.

17. On the AT&T-Fujitsu controversy see Edward Meadows, "Japan Runs into America, Inc.," *Fortune*, Mar. 22, 1982, pp. 56-61.

18. Gerschenkron, pp. 5-51.

19. See Edward Ames and Nathan Rosenberg, "Changing Technological Leadership and Industrial Growth," *Economic Journal*, Mar. 1963, pp. 13-31. The authors were among the first to point out the advantages and drawbacks of firstcomer status.

20. See Kenneth Arrow, "The Economic Implications of Learning by Doing," *Review of Economic Studies*, June 1962, pp. 155-73, and William Fellner, "Specific Implications of Learning by Doing," *Journal of Economic Theory*, August 1969, pp. 119-40.

21. Ira C. Magaziner and Thomas M. Hout, *Japanese Industrial Policy* (Berkeley: Institute of International Studies, University of California, 1980), p. 47. This paper was written before the publication of Chalmers Johnson's *MITI and the Japanese Miracle: The Growth of Industrial Policy, 1925-1975* (Stanford: Stanford University Press, 1982); it would have benefited from the historical and analytical richness of Professor Johnson's work.

22. David Apter, *The Politics of Modernization* (Chicago: University of Chicago Press, 1965). Chalmers Johnson uses the term "capitalist develop-

mental state" to describe Japan's system; the notion comes very close to the idea of a "mobilization system" in terms of power accorded the state.

23. Sakakibara Eisuke and Noguchi Yukio, "Okurasho, Nichigin Ocho no Bunseki" [An Analysis of the Ministry of Finance and Bank of Japan Dynasty], *Chuo Koron*, August 1977, pp. 96-150.

24. Kanji Haitani, *The Japanese Economic System* (Lexington, Mass.: Lexington Books, 1976), pp. 96-150.

25. Based on extensive interviews with MITI career bureaucrats.

26. Unclassified report, U.S. Embassy, Tokyo, May 1982. Dollar estimates for national research projects listed are calculated at an exchange rate of 220 yen to the dollar.

27. Interviews with MITI officials, June 1982.

28. Interview with MITI official, May 1982.

29. Interviews with Japanese semiconductor executives, June-September 1982.

30. Nakagawa Yasuzo, *Nihon no Handotai Kaihatsu* [Japan's Semiconductor Development] (Tokyo: Daiyamondosha, 1981), pp. 9-137.

31. Kagaku Gijutsu-cho, ed., *Kagaku Gijutsu Hakusho* [White Paper on Science and Technology] (Tokyo: Okurasho Insatsukyoku, 1981), p. 14.

32. See "Tomorrow's Leaders: A Survey of Japanese Technology," *The Economist*, June 19, 1982.

33. Interview with Aiko Jiro, MITI official, Apr. 14, 1982; information also derived from the mimeographed text of a speech delivered by Mr. Aiko in Washington, D.C., on Japan's efforts in high technology, Apr. 1, 1982.

34. Eugene J. Kaplan, *Japan: The Government-Business Relationship* (Washington, D.C.: U.S. Department of Commerce, 1972), pp. 86-91.

35. Nihon Denshi Keisanki Kabushiki Kaisha, ed., *JECC: Computer Note, 1982* (Tokyo: Nihon Denshi Keisanki Kaisha, 1982), pp. 401-3.

36. Nihon Joho Shori Kaihatsu Kyokai, ed., *Computer Hakusho* [Japan Computer White Paper] (1981), pp. 30-38, 68-71, 95-107.

37. For an argument in support of strict antitrust enforcement see Yasuda Osamu, "Shirarezaru Dokkin ho to Kimyo na Giron" [Puzzling Arguments about Anti-Trust Law] *Bungei Shunju*, January 1981, pp. 420-26.

38. Ishikawa Masumi, *Sengo Seiji Kozoshi* [A History of Postwar Japan's Political Structure] (Tokyo: Nihon Hyoronsha, 1978), pp. 1-178. See also Nishihira Shigeki, *Nihon no Senkyo* [Japanese Elections] (Tokyo: Shiseido, 1972).

39. Daniel I. Okimoto, "The Economics of National Defense," in Okimoto, ed., *Japan's Economy*, pp. 231-83.

40. Soma Masao, *Kokusei Senkyo To Seito Seiji* [National Elections and Party Politics] (Tokyo: Seiji Koho Center, 1977), pp. 35-94.

41. See Ryutaro Komiya, "Planning in Japan" in Morris Bornstein, ed., *Economic Planning, East and West* (Cambridge, Mass.: Ballinger, 1975), pp. 189-235.

42. On the other hand, the influence of political parties in economic management seems to be increasing for a variety of reasons. See Muramatsu Michio, *Sengo Nihon no Kanryo-sei* [Postwar Japan's Bureaucratic System] (Tokyo: Toyo Keizai, 1981), pp. 137-256.

43. I am grateful to Mr. Kenji Inaba and Mr. Takeshi Isayama, MITI officials, for sharing these insights with me, April 1982.

44. Y. Ojimi, "Basic Philosophy and Objectives of Japanese Industrial Policy," in *The Industrial Policy of Japan* (Paris: Organization for Economic Cooperation and Development, 1972), p. 15.

45. I am indebted to Professor Yukio Noguchi for pointing this out to me, June 1982.

46. See Tsuruta Toshimasa, *Sengo Nihon no Sangyo Seisaku* [Postwar Japan's Industrial Policy] (Tokyo: Nikkei Shimbunsha, 1982), pp. 159-287.

47. *The Industrial Policy of Japan*, p. 16.

48. See Ueno Hiroya, "Waga Kuni Sangyo Seisaku no Hasso to Hyoka" [The Evolution of Japanese Industrial Policy: An Evaluation], in Ueno Hiroya, ed., *Nihon no Keizai Seido* [Japan's Economic System] (Tokyo: Nihon Keizei Shimbunsha, 1978), pp. 1-118.

49. Just because Japan established the kind of industrial structure it set out to construct does not necessarily mean that its industrial policy was responsible for that success. On this point see Ryutaro Komiya, "Planning in Japan," in Bornstein, ed., pp. 189-227.

50. Chalmers Johnson attributes great significance to industrial policy in his book, *MITI and the Japanese Miracle*; Hugh Patrick and Henry Rosovsky downplay its importance and give more weight to factors in the private sector: *Asia's New Giant* (Washington, D.C.: Brookings Institution, 1976).

51. Gary R. Saxonhouse, "Japanese High Technology, Government Policy, and Evolving Comparative Advantage in Goods and Services," unpublished manuscript, May 1982.

52. See C. Fred Bergsten, "What to Do about the U.S.-Japan Economic Conflict," *Foreign Affairs*, Summer 1982, pp. 1059-75.

53. Saxonhouse, pp. 1-2.

54. Ira C. Magaziner and Robert B. Reich, *Minding America's Business* (New York: Harcourt Brace Jovanovich, 1982), pp. 241-47.

55. Yukio Noguchi, "The Government-Business Relationship in Japan," unpublished paper prepared for U.S.-Japan Economic Symposium, Honolulu, January 1981, p. 30.

56. Japan Development Bank, *Facts and Figures about the Japan Development Bank* (Tokyo, 1981), p. 26.

57. Noguchi, p. 19.

58. See James C. Abegglen, "Narrow Self-interest: Japan's Ultimate Vulnerability?" in Diane Tasca, ed., *U.S.-Japanese Economic Relations* (New York: Pergamon Press, 1980), pp. 21-31.

59. For a perceptive analysis of the French electronics industry see John Zysman, *Political Strategies for Industrial Order* (Berkeley and Los Angeles: University of California Press, 1977).

60. See Murakami Yasusuke, "Sengo Nihon no Keizai System" [Postwar Japan's Economic System], *Economist*, June 14, 1982, pp. 38-54; Ouchi Atsuyoshi, Ishii Takemochi, and Moritani Masanori, "Electronic Shock," *Voice*, February 1982, pp. 66-93.

Chapter Five

We wish to thank the other members of the U.S.-Japan Study Group for their support and their intellectual challenges in our collective search for a meaningful synthesis. We are especially grateful to Mr. Yuji Masuda, Professor Daniel I. Okimoto, Mr. W. Edward Steinmueller, and Dr. Franklin B. Weinstein. Several outside the group read our manuscript and commented on it: Professor Richard Caves, Professor Kim Clark, Professor Richard Samuels, Dr. Annette Lamonde, and Howard Bogert. We bear full responsibility for any remaining ambiguities, misinterpretations, or errors.

1. Semiconductor Industry Association, *The International Microelectronic Challenge* (Cupertino, Calif.: Semiconductor Industry Association, 1981); T. Howell, W. Davis, and J. Greenwald, *The Effects of Government Targeting on World Semiconductor Competition* (Cupertino, Calif.: Semiconductor Industry Association, 1983).

2. See, for example, B. J. Spencer and J. A. Brander, "International R&D Rivalry and Industrial Strategy," *Review of Economic Studies*, forthcoming.

3. Henry C. Wallich and Mable I. Wallich, "Banking and Finance," in *Asia's New Giant*, ed. Hugh Patrick and Henry Rosovsky (Washington, D.C.: Brookings Institution, 1976).

4. G. N. Hatsopoulos, *High Cost of Capital: Handicap of American Business* (Waltham, Mass.: Thermo Electron Corporation, 1983).

5. Peter Drucker, "Economic Realities and Enterprise Strategy," in Ezra F. Vogel, ed., *Modern Japanese Organization and Decision-making* (Berkeley: University of California Press, 1975).

6. Richard E. Caves and Masu Uekusa, *Industrial Organization in Japan* (Washington, D.C.: Brookings Institution, 1976).

7. Eisuke Sakakibara, Robert Feldman, Yuzo Harada, "Japanese Financial System in Comparative Perspective," report prepared for the Program on U.S.-Japanese Relations, Center for International Affairs, Harvard University, 1981.

8. *Electronic News*, June 29, 1981, p. 22.

9. For a general discussion of market vs. organization in resource allocation see O. E. Williamson, *Market and Hierarchies* (Free Press, 1975). For application and extension of this approach in the U.S.-Japan comparison of two economic systems see K. Imai and J. Itami, "Firms and Markets in Japan— Mutual Penetration of Market Principle and Organization Principle," *Contemporary Economics*, Summer 1981 (in Japanese).

10. W. Edward Steinmueller and Yuji Masuda, "The Role of Industry Structure in the U.S. and Japanese Semiconductor Industries" (unpublished paper prepared for U.S.-Japan Project, U.S.-Northeast Asia Forum on International Policy, Stanford University, 1983).

11. *Ibid.*   12. *Ibid.*

13. *Electronics*, Sept. 8, 1981, p. 89.   14. *Business Week*, Mar. 30, 1981.

15. M. Therese Flaherty, "Market Share, Technology Leadership, and Competition in International Semiconductor Markets," in *Research on Technological Innovation Management and Policy*, ed. Richard S. Rosenbloom (Greenwich, Conn.: JAI Press, 1983).

Chapter Six

1. I am grateful to W. Edward Steinmueller for helping me sort out and conceptualize the complicated relationships among variables and for suggesting appropriate readings from the theoretical literature on technological change. I also thank Hiroyuki Itami for constructive criticisms and comments.

2. See Richard C. Levin, "R&D Productivity in the Semiconductor Industry: Is a Slowdown Imminent?" in Herbert Fusfeld and Richard Langlois, eds., *Understanding R&D Productivity* (New York: Pergamon Press, 1982), pp. 37-54.

3. See Philip Kotler, *Principles of Marketing* (Englewood Cliffs, N.J.: Prentice-Hall, 1980).

4. See Joseph Schumpeter, *The Theory of Economic Development* (New York: Oxford University Press, 1961), chaps. 2 and 6.

5. Nakagawa Yasuzo, *Nihon no Handotai Kaihatsu* [The Development of Japanese Semiconductors]. (Tokyo: Daiyamondo, 1981), pp. 206-8.

6. Nihon Denshi Kikai Kogyokai, *IC Shuseki Kairo Guidebook, 1981* (Tokyo: Nihon Denshi Kikai Kogyokai, 1981), p. 18.

7. Nakagawa Yasuzo, *op. cit.*, pp. 96-100, 118-22. The early difficulties of perfecting manufacturing technology—the Japanese semiconductor industry's overriding catch-up objective—were emphasized by Dr. Osafune Hiroe of NEC, one of the pioneers, in an interview, June 14, 1983.

8. See Lester Thurow, *The Zero-Sum Society* (New York: Basic Books, 1980), pp. 76-102.

9. John J. Lazlo, Jr., "The Japanese Semiconductor Industry," unpublished study, Hambrecht and Quist, Jan. 21, 1982, pp. 20-21.

10. Nomura Research Institute, *SEMI Japanese Semiconductor Industry Report*, prepared for the Semiconductor Equipment and Materials Institute, Inc., 1983.

11. Interview with Hewlett-Packard researchers, March 1983.

12. See Ken-ichi Imai, "Japan's Changing Industrial Structure and the United States-Japan Industrial Relations," in Kozo Yamamura, ed., *Policy and Trade Issues of the Japanese Economy* (Seattle: University of Washington Press, 1982), pp. 61-64.

13. See Yoshio Suzuki, *Money and Banking in Contemporary Japan* (New Haven: Yale University Press, 1980).

14. Kazuo Sato, "Japan's Savings and Internal and External Macroeconomic Balance," *ibid.*, pp. 143-72.

15. Leo Esaki and Karatsu Hajime, "Shortcomings in Japan's R&D Approach," in *Japan Echo*, vol. 10, special issue, 1983, p. 29.

16. Interviews with executives from several Japanese banks based in the U.S.

17. For a stimulating discussion of market and organizational principles in Japan see Ken-ichi Imai and Hiroyuki Itami, "The Firm and Market in Japan—Mutual Penetration of the Market Principle and Organization Principle" (unpublished paper, August 1981); for a general theoretical framework see Oliver E. Williamson, *Markets and Hierarchies* (New York: Free Press, 1975), especially chapters 1, 4, 10, and 13.

18. Reinosuke Hara, *Management of R&D in Japan* (Tokyo: Institute of Comparative Culture Business Series, Sophia University, 1982).

19. John Tilton, *International Diffusion of Technology: The Case of Semiconductors* (Washington, D.C.: Brookings Institution, 1971); R. W. Wilson et al., *Innovation, Competition, and Government Policy in the Semiconductor Industry* (Lexington, Mass.: Lexington Books, 1980).

20. See Schumpeter, *Theory of Economic Development*; C. Freeman, *Economics of Industrial Innovation* (Middlesex, Eng.: Penguin, 1974); John K. Galbraith, *American Capitalism* (Cambridge, Mass.: Riverside Press, 1962).

21. National Science Foundation, *Indicators of International Trends in Technological Innovation* (Washington, D.C.: National Science Foundation, 1976); Organization for Economic Cooperation and Development, *Innovation in Small and Medium Firms* (Paris, OECD, 1982).

22. W. J. Abernathy and J. M. Utterback, *Innovation and the Evolution of Technology in the Firm* (Cambridge, Mass.: Harvard Business School, June 1976).

23. Alfred D. Chandler, Jr., *Strategy and Structure* (Cambridge, Mass.: M.I.T. Press, 1962).

24. John Jewkes, David Sawers, and Richard Stillerman, *The Sources of Invention* (New York: W. W. Norton, 1969), pp. 209-28.

25. Adlen S. Bean, Dennis D. Schiffel, and Mary E. Mogee, "The Venture Capital Market and Technological Innovation," *Research Policy* 4 (1975): 380-408.

26. Frank R. Kline, "Venture Capital and High Technology Prospects for the Future," unpublished paper, Oct. 20, 1982.

27. *Ibid.*

28. *Business Week*, Oct. 11, 1982, p. 53.

29. M. Aoki, "A Few Facts on the Ownership and Equity Structure of Large Japanese Firms," unpublished paper, November 1982.

30. *Indicators of International Trends in Technological Innovation*.

31. *Innovation in Small and Medium Firms*, p. 199.

32. Chusho Kigyo-cho, *Chusho Kigyo Hakusho, 1981* [White Paper on Small-Medium Enterprise] (Tokyo: Okurasho Insatsukyo, 1982).

33. Interview with executives of a leading Japanese electronics corporation, June 1983.

34. Michael Borrus, James Millstein, and John Zysman, *U.S.-Japanese Competition in the Semiconductor Industry* (Berkeley: Institute of International Studies, University of California, 1982), p. 60.

35. Peter Drucker, "Economic Realities and Enterprise Strategy," in Ezra F. Vogel, ed., *Modern Japanese Organization and Decision-making* (Berkeley: University of California Press, 1975), pp. 228-44.

36. Aoki, "Ownership and Equity Structure of Large Japanese Firms."

37. Yair Aharoni, *The No Risk Society* (Chatham, N.J.: Chatham House, 1981).

38. Chase Financial Policy, "U.S. and Japanese Semiconductor Industries: A Financial Comparison," June 9, 1980.

39. Itami Hiroyuki, *Nihonteki Keiei-ron o Koete* [Beyond Theories of Japanese Management] (Tokyo: Toyo Keizai Shimposha, 1982).

40. *Ibid.*, pp. 31-32.

41. *Ibid.*, pp. 34, 37.

42. Kenichi Ohmae, "The Long and Short of Japanese Planning," *Wall Street Journal*, Jan. 18, 1982.

43. *Ibid.*

44. See Eugene Kaplan, *Japan: The Government-Business Relationship* (Washington, D.C.: U.S. Department of Commerce, 1972), pp. 104-58.

45. Hugh T. Patrick and Henry Rosovsky, eds., *Asia's New Giant* (Washington, D.C.: Brookings Institution, 1976), p. 47.

46. For the early postwar period, I agree with Chalmers Johnson's emphasis on the centrality of MITI's role. See Johnson's *MITI and the Japanese Miracle* (Stanford: Stanford University Press, 1982), pp. 198-241.

47. Saito Seiichiro, "Tsusho Sangyosho-ron" [A View of MITI], *Shokun*, June 1978.

48. *The Effect of Government Targeting on World Semiconductor Competition*, prepared by Verner, Liipfert, Bernhard, and McPherson, Chartered, and copyrighted by the Semiconductor Industry Association, 1983.

49. A. M. Golding, "The Semiconductor Industry in Britain and the United States," Ph.D. dissertation, University of Sussex, 1971.

50. John Zysman, *Political Strategies for Industrial Order* (Berkeley and Los Angeles: University of California Press, 1977).

51. See, for example, William W. Lockwood, *The Economic Development of Japan* (Princeton: Princeton University Press, 1968).

52. *Effect of Government Targeting*, pp. 90-91.

53. Richard E. Caves and Ronald W. Jones, *World Trade and Payments* (Boston: Little, Brown, 1977) pp. 202-3.

54. *Effect of Government Targeting*, p. 86, n. 108.

55. See Nomura Research Institute, *SEMI Japanese Semiconductor Industry Report*.

56. Interview conducted November 1982.

57. *The Effect of Government Targeting*, pp. 43-49

58. *Ibid.*, p. 97.

59. Yukio Noguchi, "The Government-Business Relationship in Japan," in Yamamura, *Policy and Trade Issues*, pp. 123-42.

60. Ken-ichi Imai, "Japan's Changing Industrial Structure and United States-Japan Industrial Relations," *ibid.*, pp. 47-75.

61. Nico Hazewindus, with John Tooker, *The U.S. Microelectronics Industry* (New York: Pergamon Press, 1982), p. 121.

62. See Electronics Industries Association, *The Directory of Defense Electronic Products and Services* (New York: Information Clearing House, Inc., 1981), pp. iv-xiii.

63. *Ibid.*, pp. xi-xii.

64. See Kozo Yamamura, "Success that Soured: Administrative Guidance and Cartels in Japan," in Yamamura, *Policy and Trade Issues*, pp. 77-112.

65. *Effect of Government Targeting*, pp. 52-57.

66. Interviews with several U.S. semiconductor executives, November 1982.

67. Interview with MITI officials, June 1982.

68. Gary R. Saxonhouse, "Japanese High Technology, Government Policy, and Evolving Comparative Advantage in Goods and Services," unpublished paper, 1982.

69. For a balanced discussion of U.S. industrial policy see Office of Technology Assessment, *U.S. Industrial Competitiveness* (Washington, D.C.: Congress of the United States, Office of Technology Assessment, 1981), pp. 157-65.

70. For a theory of political stability based on the notion of congruent patterns see Harry Eckstein, *Division and Cohesion in Democracy* (Princeton, N.J.: Princeton University Press, 1966).

71. See, for example, Daniel Bell, *The Cultural Contradictions of Capitalism* (New York: Basic Books, 1976).

72. Samuel P. Huntington, *Political Order in Changing Societies* (New Haven: Yale University Press, 1968); Johnson, *MITI*.

73. James R. Kurth, "The Political Consequences of the Product Cycle: Industrial History and Political Outcomes," *International Organization* (Winter, 1979), pp. 1-34.

74. For a Marxist view of the relationship between big business and the state, see Robert Jessop, *The Capitalist State* (New York: New York University, 1982), pp. 32-77; and Ralph Miliband, *The State in Capitalist Society* (New York: Basic Books, 1969), pp. 55-67.

75. For a provocative theory concerning the relationship between political stability and economic growth, see Mancur Olson, *The Rise and Decline of Nations* (New Haven: Yale University Press, 1982), pp. 36-98.

76. One study suggests that the influence of political parties vis-à-vis the bureaucracies is on the rise. See Michio Muramatsu, *Sengo Nihon no Kanryosei* [Postwar Japan's Bureaucratic System] (Tokyo: Tokyo Keizai Shimposha, 1981), esp. pp. 137-68.

77. Two books that made the best-seller list in America are Richard Pascale and Anthony Athos, *The Art of Japanese Management* (New York: Simon and Schuster, 1981), and William Ouchi, *Theory Z* (Menlo Park, Calif.: Addison-Wesley, 1981); earlier studies include Rodney Clark, *The Japanese Company* (New Haven: Yale University Press, 1979); Ronald Dore, *British Factory-Japanese Factory* (London: George Allen & Unwin, 1973); Robert E. Cole, *Work, Mobility, and Participation* (Berkeley: University of California Press, 1980).

78. See, for example, Raymond Vernon, *Sovereignty at Bay* (New York: Basic Books, 1971).

79. The aircraft industry is a particularly striking example of transnational cooperation. See David C. Mowrey and Nathan Rosenberg, "The Commercial Aircraft Industry," in Richard Nelson, ed., *Government and Technical Progress* (New York: Pergamon Press, 1982), pp. 101-61.

80. To put the notion of sector-specific interdependence in broader theoretical perspective see Robert O. Keohane and Joseph S. Nye, *Power and Interdependence* (Boston: Little, Brown, 1977).

81. Tadao Kagono et al., "Japanese Inductionism vs. American Deductionism: A Comparative Analysis of Strategy, Structure, and Process," unpublished paper, March 1983, p. 2.

Index

Acquisitions, capital, 195-96
Administrative mechanisms, 112-13, 147f, 151, 158. *See also* Macroeconomic management
Advanced Micro Devices (AMD), 58-59, 74, 90n, 95, 142, 161, 185, 191
Agency for Industrial Science and Technology (MITI's), 19, 21, 102, 194
Agriculture, 128-29, 213n
Aircraft industry, 266
Air Force, U.S., 79-80
Albuquerque, New Mexico, 172
Alpha particle problems, 52-53
AMD, *see* Advanced Micro Devices
Amdahl, 44, 49, 75
American Business Conferences, Inc., 138
American Electronics Association (AEA), 29-32, 60, 121, 188n
American Microsystems, Inc. (AMI), 16, 73, 209-10
American Telephone and Telegraph (AT&T), 3, 16, 56, 73, 79, 93, 154-64 *passim. See also* Bell Laboratories
Ames, Edward, 259
AMEX, 153
Anderson, Richard W., 53
Ando Electric, 184
Anelva, 184n
Antitrust policies, 100, 111-14, 187
Apollo spacecraft, 84, 108
Applied Materials, Inc., 258
Apter, D., 97
Arizona, 74
Army, U.S., 79-80. *See also* U.S. military programs
Ashton, Peter K., 84, 239

AT&T, *see* American Telephone and Telegraph
Austin, Texas, 13, 172
Automation, 48, 53, 54-55, 60-69 *passim*, 76-77
Automobile industry, 76-77, 92, 231

Banking-industrial complex, Japanese, 4, 7, 151-52, 158, 199f, 211, 234
Bank of America, 151
Bank of Japan, 7
Banks: Japanese, 7, 93, 100, 110, 126, 130-33 *passim*, 140-58 *passim*, 175, 186, 191-205 *passim*, 211f, 222, 233; U.S., 7, 141-45 *passim*, 150f
Bell Laboratories, 9f, 14f, 48f, 80, 106, 190; training by, 34, 79; and alpha particle problems, 52; antitrust policies and, 113
Bipolar gate arrays, 49
Bipolar ICs, 14-15
Bomar, 16
Bonding machines, 62-67 *passim*
Bonds, 138-49 *passim*, 156, 158, 211
Boston Consulting Group, 73
Burroughs, 44
Busicom, 15, 16-17

Calculators, 15, 16-17, 89, 179
California, semiconductor industry in, 42, 60, 66f, 75, 120, 129, 154, 190f
California, University of, 12
Canon, 184
Capital acquisitions, 195-96
Capital availability, 6-7, 25, 70, 87, 126-64 *passim*, 174-76, 191-213 *pas-*

sim, 220-34 *passim. See also* Loans; Venture capital

Capital costs, *see* Costs

Capital expenditures, 7, 70, 165-73, 185-87, 197-98, 227, 229, 233, 258

Capitalist development state, 259-60

Cartelization, 113, 187, 222

Casio, 133

Catch-up, 95-115 *passim*, 182, 193, 217, 227, 263

Caves, Richard E., 146, 262

Chase Financial Policy study, 203-8 *passim*

Chase Manhattan Bank, 203

Chrysler, 1, 122, 203

Clark, Kim, 262

Components production, vs. systems production, 72-73

Computer-aided design (CAD), 48, 187

Computer Development Laboratories (CDL), 18f

Computer industry, 17-22 *passim*, 50, 72-73, 93-94, 109-11, 156f. *See also* IBM

Computer science training, 30-33 *passim*, 48, 240, 242, 246-47

Congress, U.S., 92-93, 113, 116

Consumer electronics market, 21f, 24, 88-89, 179, 235

Cornell University, 12

Corrigan, Wilfred, 74

Costs: capital, 7, 25, 70, 134, 138-40, 173-76 *passim*, 203-11 *passim*, 233-34; labor, 63-64; automation, 66; R&D, 82, 86, 95. *See also* Capital expenditures; Prices; Salaries

Courts, 113

Cross-investment, 74-77

Cross-licensing, 27

Cultural factors: in quality control, 61-62; in venture capital, 153-54; in policymaking, 218

Customized chips, 46-47, 48, 73-74, 75

Dataquest, 74, 163, 172f

Debt/equity ratios, 140f, 147-48, 149, 198-202 *passim*

Debt financing, 144, 198-212 *passim*, 234

Defense, Japanese, 95-96. *See also* U.S. military programs

Defense Advanced Research Projects Agency (DARPA), 11-12, 37, 225

Demand, for capital, 139-40

Demand pull, 83, 100, 108-11, 196. *See also* Market demand

Deming, William, 5

Democratic Party, 116

Department of Agriculture, U.S., 213n

Department of Defense, U.S., 10, 55-56, 184; VHSIC program of, 11, 24, 37, 113, 154-55, 183; supercomputer project of, 11, 37, 94-95, 225; and transistors, 23, 86; and trade balance, 93; and process technology, 182-83. *See also* U.S. military programs

Department of Justice, U.S., 94

Design automation (DA), 48

Die-attachment machines, 66-67

Diet, Japanese, 112-20 *passim*

Digital Equipment, 162

Direct external financing, 141, 145f, 212. *See also* Bonds; Equity capital

Displays, 44

DOD, *see* Department of Defense

Drucker, Peter, 198

Dual economy, 139

Dumping, 15, 54, 90n, 220, 223-24

Dunn, Roger, 53-54, 68

Duran, J. M., 5

Eagleton, Thomas, 92n

Earnings, retained, 138, 142, 148, 157, 162-64

Economic Recovery Tax Act, 12

Economy: Japanese, 4, 98, 116, 126, 139, 158, 161n, 212-15 *passim*; U.S., 116-17, 127, 161n, 186, 213, 226, 228. *See also* Financial systems; Macroeconomic management; Market demand; Recession

Education, pre-university, 48, 61. *See also* Training

Egan, Thomas P., 84, 239

Electoral systems, 115-19, 229

Electrical Communication Research Laboratories (of NTT), 18, 47

Electron beam lithography, 10, 51, 183

Electronics, 168

Electrotechnical Laboratory, 19, 21, 102

Employees: U.S., 31f, 59-67 *passim*, 120, 153, 188-89; salaries of, 31f, 188-89; turnover of, 59-61, 187-88, 189; Japanese, 59-64 *passim*, 153, 187-89, 192, 209. *See also* Engineers

End-user demand, 84, 89, 111, 179-80, 235. *See also* Market demand

Engineers, training of, 12, 28-34, 48, 60f, 79, 95, 240-47
Entrepreneurs, 9-10, 77, 113, 152-54, 191-93, 232. *See also* Venture firms
Equity capital, 138, 141; of Japanese banks, 7, 144-46, 199, 202, 204; for Japanese firms, 7, 143-49 *passim*, 153, 199-207 *passim*, 211ff; for American firms, 142-58 *passim*, 195-99 *passim*, 204, 207f
Esaki, Leo, 187
Europe, 27, 105, 108f, 114, 203, 227. *See also* France; United Kingdom
Exchange rate, 127-32 *passim*, 161n, 207, 260
Expenditures, *see* Capital expenditures; Costs
Export deluge, from Japan, 91, 93, 105-6, 217, 222-23
External financing, 138-59 *passim*, 211f, 233. *See also* Bonds; Equity capital; Loans
Exxon Foundation, 32

Fairchild, 15, 73f, 142, 160-64 *passim*; government-sponsored R&D at, 23, 81; Japanese plants of, 28, 75; capital expenditures of, 167, 169, 172
Fairchild Camera and Instrument, 93
Fair Trade Commission (FTC), 111, 114
Federal Reserve, U.S., 127
Fifth-Generation Computer Project, 21, 93ff, 216, 221, 225, 227
Finances, 12-13, 33-34, 37, 77, 134-76, 197-213, 233-34. *See also* Capital availability; Cross-investment; Government funding; Sales
Financial systems, 7, 135-59, 174ff, 197-213, 226-27. *See also* Banks; Taxes
Firstcomers, 90, 95f, 108, 259. *See also* Pioneers
Fiscal Investment and Loan Program (FILP), 130, 132, 220
Ford, 1, 74
Four-Phase Systems, 72
France, 107, 135, 193, 216f, 221, 225
Frieden (H. P.), 16
FTC, *see* Fair Trade Commission
Fujitsu, 39-49 *passim*, 75, 93, 161, 177, 184, 218, 230; in VLSI projects, 18; and loan financing, 144; and JECC, 156; NTT and, 157; capital expenditures of, 167-72 *passim*
Fukuda faction, 120

Galbraith, John K., 190
Gallium arsenide technology, 21, 44-45, 69-70
Gates, described, 46
General Accounts Budget (GAB), Japanese, 118, 132
General Agreement on Tariffs and Trade (GATT), 8, 91, 105, 220
General Electric, 14, 80
General Microelectronics (GME), 16
Germanium transistors, 14
Gerschenkron, Alexander, 95, 259
Gilpin, Robert, 88
Glass-Steagull Act, 7
Gould, 73
Government funding: Japanese, 21-24, 34, 49, 94, 96-97, 102-8 *passim*, 117, 129-35 *passim*, 155-57, 176, 183, 194-95, 220-22; U.S., 22-24, 33-34, 37, 79-95 *passim*, 106f, 135, 154-55, 176, 182, 221; of R&D, 22-24, 33-34, 37, 79-83, 95, 102-8 *passim*, 182f, 220; of software, 49. *See also* Procurements, public
Governments, U.S. state, 129. *See also* Japanese government; U.S. government
Grand coalition, LDP's, 117-19
Great Britain, 2, 107, 193, 216f, 221
Growth, in semiconductor industry, 76, 77, 121, 172, 211-12, 214

Hambrecht and Quist, 71
Hatsopoulos, G. N., 138-39
Hayakawa, 15f
Hazewindus, Nico, 180
Hewlett-Packard, 5, 49, 53-62 *passim*, 68, 74, 162, 184f, 210
High electron mobility transistors (HEMTs), 44
Hiroe, Osafune, 263
Hitachi, 39-44 *passim*, 50f, 161, 184, 218, 230; MOS production by, 15; in VLSI projects, 18; automation at, 63, 65; cross-investment by, 74f; and loan financing, 144; and JECC, 156; NTT and, 157; capital expenditures of, 167-70 *passim*
Hitachi America, Ltd., 75
Hoenni, Jean, 10
Hoff, Ted, 10
Hughes, 10

IBM, 25, 36-48 *passim*, 154-64 *passim*, 177, 190, 210, 227; electron beam

lithography at, 10; MOS production at, 15f; and technology transfer, 28; and quality, 52, 56, 58; automation by, 64; components purchased by, 73; cross-investment by, 75; JECC and, 110f, 156; Intel holdings of, 144, 195-96
Imports, to Japan, 91, 105, 217
Inaba, Minoru, 156
Indirect external financing, 141-51 passim, 186, 211f, 233. *See also* Loans
Indirect policies, 115-33
Industrial Bank of Japan, 130
Industrial policy, 6, 119-31 passim, 216, 220-30, 234; targeted, 97n, 101-6, 155, 215, 220, 224, 227; and financing, 100-110 passim, 126-31 passim, 155, 194, 199f, 203, 220-22, 229; and industrial structure, 193-94, 227, 261
Industrial Products and Services, 160
Industrial Science and Technology Agency (ISTA), 19, 21, 102, 194
Industrial structure, 24-25, 35, 113, 189-96, 227, 261
Industrial Structure Deliberation Advisory Council, 101
Industry-university cooperation, 31-34 passim, 95
Inflation, 127, 161n, 186, 226
Information Industries Caucus, 119-20
Innovation: technological, 2, 5, 9-18 passim, 50-53 passim, 63f, 76, 107, 113, 117-97, 227, 232; organizational, 20; training, 34. *See also* Product development
Institute of Electrical and Electronics Engineers (IEEE), 31
Insurance companies, 140, 146, 156, 158
Integrated circuits (ICs): described, 2; invention of, 10
Intel, 95, 161; microprocessors at, 10, 15, 17, 49f, 68; MOS devices at, 16; and technology transfer, 28; RAMs at, 40-43 passim, 70; and alpha particle problem, 52; quality control at, 58, 68; computer systems at, 72-73; custom chips at, 73-74; financing of, 144, 191, 195-96; capital expenditures of, 167-73 passim, 185-86; venture group of, 193
Interest groups, 117-21, 229
Interest rates, 100, 127, 130, 137, 149, 199n, 205, 234
Internal consumption, 159-62 passim, 184

Internal financing, 138-48 passim, 157-64 passim, 175, 212
International Business Machines, *see* IBM
International Solid State Circuits Conferences, 27
International Trade Commission (ITC), 218
Investment push, 196. *See also* Technology push
Investments, *see* Capital availability; Capital expenditures; Cross-investment
Ion-beam machine, 51
Ion implantation, 51, 62
IPL Systems, 75
Itami, Hiroyuki, 208-11

Japan Development Bank (JDB), 93, 100, 110, 130-31, 133, 155ff
Japan Electronic Computer Corporation (JECC), 109-10, 111, 156
Japanese government, 5-6, 18-24 passim, 33, 45, 94-137 passim, 176, 183, 194-205 passim, 213-34 passim; and loan guarantees / signaling mechanisms, 7, 130-31, 148, 155, 198-203 passim, 222; funding by, 21-24, 34, 49, 94, 96-97, 102-8 passim, 117, 129-35 passim, 155-57, 176, 183, 194-95, 220-22; taxes of, 25, 67, 99-100, 128, 153, 226. *See also* Industrial policy; Ministry of Finance; Ministry of International Trade and Industry
Japan Export and Import Bank, 148
Japan Research and Development Corporation, 194
Johnson, Chalmers, 259-60, 261, 265
Josephson junction technology, 44f
Juran, J. M., 51

Kagono, Tadao, 182n
Kangyo Denki Kikai, 191n
Keiretsu membership, 146, 150
Kilby, Jack, 10
Kim, K. H., 257
Kleinman, H., 88
Knowledge-Intensive Industries Caucus, 119-20
Kokusai Electric, 184
Kurlih, Thomas, 71
Kyushu, 172

Labor, *see* Employees
Large scale integration (LSI) design, 15, 68, 246

Lasers, long wave length semiconductor, 44
Latecomers, 85, 95-96, 108, 217, 259
Lazlo, John J., Jr., 71
Learning curve, 73, 86-87, 181, 233, 256
Legislation, 7, 92-93, 100f, 111-14, 187. *See also* Taxes
Liberal-Democratic Party (LDP), 115-20 *passim*, 132, 213, 229
Licensing, 27
Linear ICs, 179-80
Loans, 138-58 *passim*; of Japanese banks, 7, 93, 100, 110, 126, 130-31, 133, 141-58 *passim*, 191, 198-201 *passim*, 212, 222; of U.S. government, 122
Lockheed, 122, 203
Logic devices, 36-37, 46-50 *passim*, 69-75 *passim*. *See also* Microprocessors
LSI design, 15, 68, 246

Macroeconomic management, 88, 131-32, 136f, 176, 186, 206f, 235
Malaysia, 64
Management, corporate, 67-68, 208-10, 230
Market demand, 43, 196, 235; in Japan, 21-22, 88-92, 100, 108-11, 179-85, 197; in U.S., 21-22, 83-92, 96, 108, 179, 182-83, 197
Market mechanisms, 98-99, 111, 122, 147, 158, 216
Market share, 111; Japanese, 37-42 *passim*, 50, 63, 70-72, 90-92, 106, 133, 172, 197f, 210-11, 233; U.S., 39-42, 70-73 *passim*, 90, 133, 172, 209-10, 233
Masuda, Yuji, 159f, 262
Mathematics education, 61
Matsushita, 1, 110, 133, 159ff, 218; RAMs of, 43; microprocessors of, 50; cross-investment of, 75; financing of, 144, 201, 212; capital expenditures of, 167, 170
MCC, *see* Microelectronics and Computer Technology Corporation
Memorex, 151
Memory devices, 36-37, 50, 53, 70-72. *See also* RAMs
Mergers, 195f
Metal oxide semiconductor (MOS), 15-17, 86, 91, 179-80
Microelectronics and Computer Technology Corporation (MCC), 13-14, 94, 187, 225

Microelectronics Center of North Carolina, 13
Microprocessors, 4, 10, 15, 17, 46-50 *passim*, 68, 72f, 257
Military, Japanese, 95-96, 98. *See also* U.S. military programs
Military-industrial complex, U.S., 4, 97, 151
Ministry of Agriculture and Forestry, Japanese, 118
Ministry of Construction, Japanese, 118
Ministry of Education, Japanese, 34
Ministry of Finance (MOF), Japanese, 7, 98, 119f, 136-37, 149, 152, 203, 205
Ministry of International Trade and Industry (MITI), Japanese, 17-28 *passim*, 50, 87, 93-107 *passim*, 183, 213-25 *passim*, 234, 265; software program of, 20, 47; and financing, 23, 107, 126, 140-41, 155f, 175, 192, 194, 203, 205, 221; and antitrust, 112-13; LDP and, 119f; and industrial structure, 192, 194. *See also* VLSI project
Ministry of Post and Telecommunication, Japanese, 18
Minnesota, University of, 12
Minuteman missile program, 10, 23, 84, 108
M.I.T., 13, 34, 80, 240
MITI, *see* Ministry of International Trade and Industry
Mitsubishi, 161f; in VLSI project, 18; RAMs of, 42, 169; gallium arsenide projects at, 44; Josephson junction projects at, 44; process innovation at, 51; cross-investment by, 75; loan financing of, 144; and JECC, 156; capital expenditures of, 169
Mitsubishi Bank, 150
Mitsubishi Electric, 133
Mitsubishi Trust and Banking, 146
Mobilization state, 97-98, 260
MOF, *see* Ministry of Finance
Morita, Akio, 67
MOS, *see* Metal oxide semiconductor
Mostek, 142, 160f; calculator components from, 16f; RAMs of, 40, 42, 72, 169; United Technologies' acquiring of, 73, 160f; financing of, 144, 191
Motorola, 49, 72, 90n, 142, 160f, 230; and technology transfer, 28; RAMs of, 39f, 75; cross-investment by, 74f; financing of, 81, 155; capital expenditures of, 167-72 *passim*, 186
Mowrey, David C., 180n

Multinationalization, 108, 231
Musashino Electrical Communications
  Laboratories, 42, 44, 102

NASDAQ, 153
National Advanced Systems, 75
National Aeronautics and Space Admin-
  istration (NASA), 79-80
Nationalization, 216
National Science Foundation, 4, 193
National security, 3, 39, 93f, 98. *See also*
  U.S. military programs
National Semiconductor (NS), 160f; transistors of, 15; RAMs of, 41, 43; quality
  control at, 58; employee turnover at,
  60; and computer systems, 72; custom
  chips at, 74; cross-investment in, 75;
  financing of, 144; capital expenditures
  of, 167-73 *passim*
NEC, 50f, 133, 157, 161f, 218, 230; and
  MOS circuits, 15f; RAMs of, 39, 41f;
  gallium arsenide projects of, 44;
  Josephson junction projects of, 44;
  automation at, 66; and quality, 68;
  and cross-investment, 74f; financing
  of, 144-51 *passim*, 156f, 201; internal
  consumption by, 159, 184; capital expenditures of, 167-71 *passim*
NEC-Toshiba Information Systems
  (NTIS), 18-19
Nikon, 184
Nippon Denso, 162
Nippon Electric Company, *see* NEC
Nippon Electronic Engineering Co., 164n
Nippon Steel, 1
Nippon Telegraph and Telephone (NTT),
  23-24, 44, 164n; and VLSI projects,
  18, 42, 102; procurements by, 18, 92,
  100, 109f, 157; RAMs of, 41f, 52,
  55; and software, 47; microprocessors
  of, 49; and quality, 52, 55f; and Oki,
  169n
Nissan, 1
Nitride passivation technology, 52-53
Noguchi, Yukio, 98, 131
Nomura Research Institute, 163
Nonaka, Ikujiro, 182n
North American, 16
Noyce, Robert, 10ff
NTT, *see* Nippon Telegraph and
  Telephone

Ohmae, Kenichi, 210
Oil crisis, 211-12

Ojimi, Y., 126
Oki, 28, 41f, 110, 143, 156-61 *passim*,
  169, 172
Oki Univac Kashi, 75
Okumura, Akihiro, 182n
Optical fibers, 44
Optoelectronic ICs (OEICs), 21, 28, 38,
  44, 69-70
Original equipment manufacturer (OEM)
  demand, 180
Overseas assembly facilities, U.S., 63-66, 67

Palo Alto, California, 75
Patrick, Hugh T., 214, 261
Pennsylvania, University of, 80
Perkin Elmer, 64
Petroleum industry, 113
Philips Corporation, 73, 160f, 163, 175
Pioneers, 78-95, 96, 108, 133
Plasma etching, 51
Political factors, 78-133, 228-30
Political parties, 112-20 *passim*, 132,
  213, 229, 260, 266
Preventive quality control, 56-57
Prices: cutting of, 39, 41, 43, 54, 72f, 90,
  128, 220, 223, 256f; learning curve,
  73, 256; public procurements and, 86,
  88, 110; fixing of, 111, 113; and capi-
  tal costs, 173-74, 208
Process technology, 36, 37-38, 48-69
  passim, 76-77, 229, 263; innovation
  in, 5, 9f, 50-53 *passim*, 63f, 177-85
Procurements, public, 18-22 *passim*, 79,
  83-92, 100, 108ff, 154, 157
Product development, 4, 36-52 *passim*,
  69, 86, 177-83 *passim*, 191
Production technology, *see* Process
  technology
Production volume, 3, 76, 86-87, 91,
  179, 181
Product portfolios/strategies, 208-10
Professionals, *see* Engineers
Profits, 73, 174f, 198, 202, 208, 210-11,
  223, 233
Protectionism, 93-94, 100, 105-6, 122-
  29 *passim*, 217, 231
Pursuers, 90, 95-115, 133. *See also*
  Catch-up

Quality, 5, 50-69 *passim*, 76-77, 185

RAMs (random access memory), 70-72;
  16K, 5, 39f, 51-55, 67-72 *passim*, 223,
  257; 64K, 3, 36-43 *passim*, 67-72 *pas-*

sim, 105-6, 168-69, 223; 256K, 3, 36, 39-43, 71
R&D, *see* Research and development
Raytheon, 80
RCA, 14f, 28, 80f
RCA Service Company, 75
Reagan administration, 121
Recession, 43, 69f, 163, 167, 173, 205, 222
Redundancy principle, 41
Republican Party, 116
Research and development (R&D), 196; U.S., 4, 10-14, 22-28 *passim*, 33-38 *passim*, 47, 78-83, 94-95, 106f, 113, 154-55, 173, 182-87 *passim*, 227; Japanese, 4, 17-28 *passim*, 34-38 *passim*, 75, 100-108, 173, 183, 193, 217-22 *passim*, 227; joint U.S.-Japanese, 27-28, 231. *See also* Innovation; VLSI project
Research and Development Project of Basic Technology for Future Industries, 107-8
Reserve funds, 201
Return on investment (ROI) strategies, 191, 207-11 *passim*, 223, 227, 233
Rockwell, 16
Rolm Corporation, 196
Rosen, Benjamin, 53, 71, 77
Rosenberg, Nathan, 180n, 259
Roseville, California, 42, 66, 75
Rosovsky, Henry, 214, 261
Royama, Shoichi, 141

Sakakibara, Eisuke, 98
Sakakibara, Kyonori, 182n
Salaries, 31ff, 188-89
Sales, 41, 89, 159-64 *passim*, 186. *See also* Market share
Sales/capital expenditure ratio, 165, 169-73, 185-86, 197-98, 258
Samuels, Richard, 262
Sanders, W. J., 90n
San Diego, California, 67, 75
San Francisco Bay Area, 60
Santa Clara, California, 75
Sanyo Electric, 143
Savings, 130, 137, 139f, 149, 158, 186, 206
Schlumberger, 73, 160-64 *passim*, 169n, 175
Schumpeter, Joseph, 190
Science education, 61
Seiko, 133, 162

Semiconductor Industry Association (SIA), 12, 120f; on trade issues, 40, 92, 215, 220, 223, 256f; on finance issues, 134, 165
Semiconductor Research Cooperative (SRC), 12ff
Separation, of financial systems, 137, 140, 174
Sharp, 15, 133, 144, 161
Shea, John, 257
Shinkawa, 184
Shugart, 193
SIA, *see* Semiconductor Industry Association
Signetics, 73, 142, 160f, 163, 172f
Silicon foundries, 73-74
Silicon-on-sapphire (SOS) technology, 44
Silicon transistors, 14
Silicon Valley, 120, 154, 190f
Size, company, 25, 113, 189-95, 227
Small Business Finance Corporation, 194
Small-Medium Enterprise Agency, 194
Software, 4-5, 20, 46, 47-49, 69, 72-73, 257
Sony, 75, 133, 159, 161ff, 212; transistors of, 14; tunnel diode invention at, 37; management at, 67; financing of, 146, 147-48
Southeast Asia, U.S. plants in, 63, 64-65
Space programs, U.S., 10, 22f, 29, 55-56, 79, 84, 108
Sperry, 28, 75
Sperry-Univac, 44
Stanford University, 12, 34, 80
State governments, U.S., 129
Steinmueller, W. Edward, 159f, 262f
Stocks, *see* Equity capital
Subsidies Code, 220
Sumitomo, 146, 150
Sunnyvale, California, 75
Supercomputer project, 11, 37, 94-95, 225
Sylvania, 1, 80
Systems production, vs. components production, 72-73

Takeda Riken, 184
Takeovers, 195
Tanaka faction, 120
Targeted industrial policies, 97n, 101-6, 155, 215, 220, 224, 227
Tariff rates, 109
Taxes: in U.S., 12, 37, 116, 121f, 128, 153, 182, 186, 191, 226; in Japan, 25, 67, 99-100, 128, 153, 226

Technology, 2-77 *passim*, 227, 229, 232. *See also under* Innovation

Technology push, 80, 83, 96, 101, 108, 196

Technology transfer, 16, 25-28, 37-38

Telecommunications industry, 9f, 17f, 23-24, 56, 157, 196. *See also* American Telephone and Telegraph; Nippon Telegraph and Telephone

Television instruction, 34

Temporary Law for the Promotion of Specific Machinery and Information Industries, 101

Texas, 13, 75, 172f

Texas Instruments (TI), 161, 190, 227, 230; electron beam lithography at, 10; transistors of, 14-15; MOS circuits of, 16; and technology transfer, 28; memory devices of, 39-40, 42, 70, 75; and quality, 52; automation at, 64f; pricing strategies of, 73; cross-investment by, 75; financing of, 80-81, 142, 144, 155; capital expenditures of, 167-71 *passim*, 185-86

Thurow, Lester, 182

Tilton, John E., 81, 85

Tokuda Seisakusho, 184n

Tokyo, University of, 33, 44, 48, 240-47

Tokyo High Court, 113

Tokyo Institute of Physical and Chemical Research, 44

Tokyo Rounds, 220

Tokyo Sanyo, 143, 161

Tooker, Gary, 72

Toshiba, 47, 75, 133, 161f, 218; in VLSI project, 18-19; RAMs of, 39, 43; ion-beam machine of, 51; automation at, 62-63; financing of, 144; and JECC, 156; capital expenditures of, 169f

Total Quality Control (TQC), 56

Toyota, 1

Trade, international, 8, 91-94, 98, 216-23 *passim*, 227-31 *passim*; corporate finance structure and, 93, 122, 135, 145, 222; targeting and, 104-6, 220; exchange rate and, 128, 131, 207

Trade Act, 93

Training, of professionals, 12, 28-34, 48, 60f, 79, 95, 240-47

Transistors, 14-15, 23, 44, 79, 88

Transnationalization, 7-8, 27-28, 74-77, 231-32, 266

TRW, 75

Tsongas, Paul, 113n

Tufte, Edward R., 117n

Tunnel diode invention, 37

Turnover, labor, 59-61, 187-88, 189

Uekusa, Masu, 146

Ulvac, 184n

United Kingdom, 2, 107, 193, 216f, 221

U.S. government, 5, 78-95, 101, 113-36 *passim*, 156, 176, 213-28 *passim*; space program of, 10, 22f, 29, 55-56, 79, 84, 108; taxes of, 12, 37, 116, 121f, 128, 153, 182, 186, 191, 226; funding by, 22-24, 33-34, 37, 79-95 *passim*, 106f, 135, 154-55, 176, 182, 221. *See also* Department of Defense

U.S. military programs, 4, 10-12, 21-27 *passim*, 97, 101, 123, 151, 154, 179-84 *passim*, 221; Minuteman missile, 10, 23, 84, 108; VHSIC, 11, 24, 37, 113, 154-55, 183; supercomputer project in, 11, 37, 94-95, 225; and quality, 55-56; and process technology, 182-83

U.S. Steel, 1

United States Trade Representative (USTR) Office, 92

United Technologies, 73, 160f, 175

Univac, 85

University research, 11-13, 20, 37, 80, 95

University training, 29-30, 31-34, 48, 61, 95, 240-47

Value added, 89-90, 208f, 223, 230

Venture capital, 25, 49, 140, 148, 152-54, 175-76, 191-92, 227, 232

Venture firms, 25, 113, 152-54, 175-76, 191-93

Vertical integration, 184, 195, 208

Very High Speed Integrated Circuits (VHSIC) program, 11, 24, 37, 113, 154-55, 183

Very Large Scale Integration (VLSI) R&D, 11, 15, 25, 39, 189, 246. *See also* VLSI project

Vietnam war, 29

Virtuous cycle, 181

VLSI Design, 47-48

VLSI project (Technology Research Association), 17-28 *passim*, 38, 113f, 155, 217, 227; RAM development of, 42; process innovation of, 50-51, 183; political caucuses and, 120

Wall Street Journal, 90n
Werner, Jerry, 47-48
Western Electric, 25; RAMs of, 40, 43; and AT&T, 56, 93, 157; finances of, 80, 83; antitrust policies and, 113
West Germany, 27
Westinghouse, 81

Wilson, Robert W., 84, 239
World War II, 79

Xerox, 54, 68

Zenith, 1